LISTEN
TO THE
WOOL

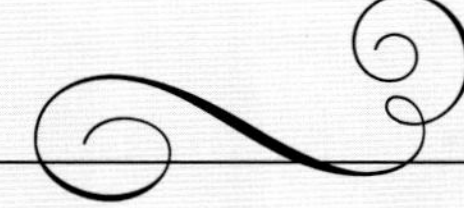

In loving memory of my father

and

To Cecilia,
Who is always the reader in my mind

LISTEN TO THE WOOL

A Why-to Guide for Joyful Spinning

JOSEFIN WALTIN
Photography by Dan Waltin

STACKPOLE BOOKS
Essex, Connecticut

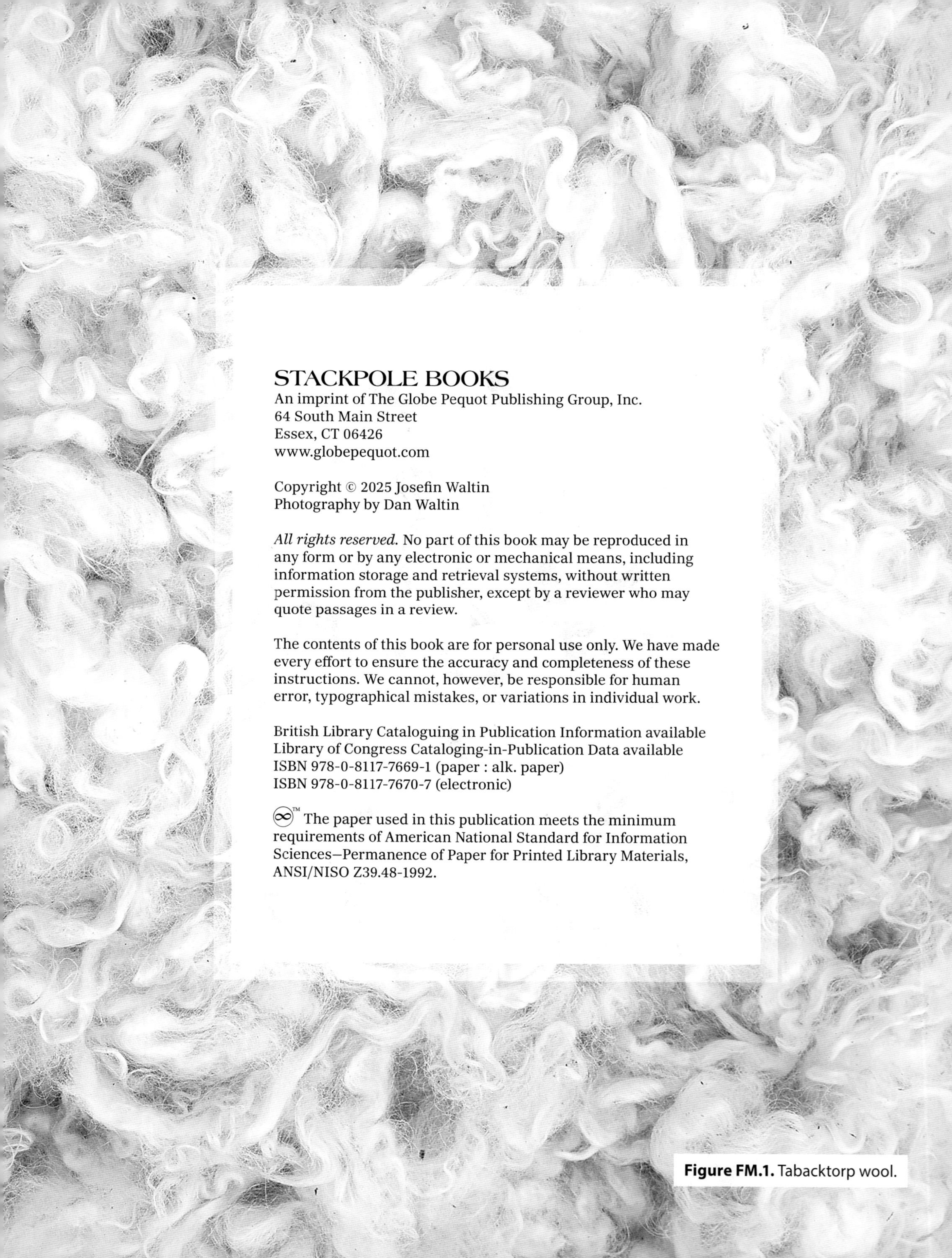

STACKPOLE BOOKS
An imprint of The Globe Pequot Publishing Group, Inc.
64 South Main Street
Essex, CT 06426
www.globepequot.com

British Library Cataloguing in Publication Information available
Library of Congress Cataloging-in-Publication Data available
ISBN 978-0-8117-7669-1 (paper : alk. paper)
ISBN 978-0-8117-7670-7 (electronic)

∞™ The paper used in this publication meets the minimum requirements of American National Standard for Information Sciences—Permanence of Paper for Printed Library Materials, ANSI/NISO Z39.48-1992.

Figure FM.1. Tabacktorp wool.

CONTENTS

FOREWORD

Quite a few books have been written to present the hows of spinning, including the details of handling fresh wool. Not so many consider the whys, and I'm not aware of another that does so in as wide-ranging and comprehensive a manner as this volume.

In twenty-first-century industrialized countries, a small number of people choose to learn to spin yarn. (It seems like there are a lot of us, and by certain measures there are, but the illusion comes from the fact that we all hang out in the same places, online and in real life.) An even smaller number bypass the delectable prepared fibers offered to us and seek the raw fleece to begin our practice, whether from the sheep in the field (the best) or from its caretaker at a festival or online.

As Josefin Waltin explains, engaging in this anachronistic craft, especially if we start with fiber grown by an animal we have met who is tended by a shepherd or rancher we know, offers all sorts of benefits difficult to obtain through any other means. Working with our hands to move local resources from potential utility to actual product—that is, taking the sheep's fleece through the transformation from the animal's protective covering to a garment for ourselves, or a rug, or a tapestry—satisfies needs far beyond the practical.

The fleece is unique. The spinner is an individual. The interaction between the two will be unrepeatable. Each time a new connection occurs, however, it will build on the foundation of previous experiences.

Josefin begins with the idea of locating our nearby fiber sources, using as examples the breeds of sheep to which she has access. Beyond that, she illuminates the lessons the fiber can teach us, including the ways it can connect us to ourselves, as well as our environment.

Mechanical processing diminishes the fibers and reduces our connection to the natural world—something modern societies seem intent on doing in many ways. A subtext here involves detaching from the digital world (which may yet supply instructional resources, although the actions we take are necessarily tactile instead).

When we discover local, distinct, fresh sources of wool, we set in motion a journey through which we produce something that is unlike anything that has been made before. At the same time, we tease forward our own skill and knowledge. Even experienced spinners find that using a new method or approach requires what in Zen practice is called "beginner's mind," the humble attitude from which growth occurs. Each technique or device instructs our fingers, minds, and spirits anew.

I've spent decades exploring the idiosyncratic qualities of a multitude of fleeces, along with different tools and techniques. I know very well the joy of pulling a lock out of a newly shorn fleece and gauging its length, crimp, color, and texture—wondering which adventure it will lead me to experience. Josefin's contributions to our community encourage us to open our hearts and our hands to the incomparable pleasures offered by every woolly sheep, who annually shares its bounty with us and bounces back into the field, already growing the fleece that will carry subtle traces of the year to come into our future crafting quests.

I encourage you to open this book and your mind and let Josefin guide you on this incomparable voyage.

Deborah Robson,
coauthor of *The Fleece & Fiber Sourcebook*

PROLOGUE

> For me, writing is an act of reciprocity with the world; it is what I can give back in return for everything that has been given to me. And now there is another layer of responsibility, writing on a thin sheet of tree, and hoping the words are worth it. Such a thought could make a person set down her pen.
>
> —Robin Wall Kimmerer, *Braiding Sweetgrass*, 152

The book you hold in your hands is a gift. You or someone who gave it to you may have paid for it, but the words and the dedication behind them are my gift to you. I hope this book is beneficial for you, both in guiding you on your spinning path and in introducing new ways of experiencing spinning, allowing them to land softly in your heart. As I listen to the wool, it tells me its secrets. Spinning the yarn as beautifully as I can is my way of honoring the wool, of giving back for all that it has taught me. Sharing what I learn with the world is a way of paying forward.

I want to feel connected to the pastures the sheep has grazed. I want to meet the wool as the protective shield it once was to the sheep. When I see a seed, a needle, or piece of grass in the fleece, when I realize what side the sheep has been sleeping on, I am reminded of the wool as a gift from a living being. Every time the wool goes through my hands, I change it a little on its journey to a protection for me. As I go on my own journey through the wool, a whisper from the sheep travels with me. When I wear a garment I have spun from a fleece, I dress myself in all that I have learned—every experience, every mistake, every minute I have spent with the wool is forever entwined in that garment. Isn't it my responsibility, then, to use this gift as mindfully as I can, making the most of what I have received, and give something back to the world?

The gift of wool moves back and forth, from the sheep, through my hands and out into yarn, to gratitude for what I have learned. There is a dynamic value in the wool I have known from that newly shorn fleece. An industrial preparation is of course still wool, but to me it has a static value as a commodity. I still pay for the fleece I buy, but it gives me other dimensions than an industrially processed material ever would. I have a responsibility to reciprocate that gift, back as well as forward. By listening to the wool, by making the best yarn I can, by letting the soul of the fleece shine in the yarn I create. By letting the sheep owner know how I have cherished the soul of the wool and made something beautiful out of their care and work. By sharing my experiences with other spinners. By writing this book.

This isn't a book of yarn recipes. The sheep already made a wool cake from its own special ingredients, and in a unique combination. With the guidance on these pages, you will be able to deconstruct the wool cake, uncover its ingredients, and make up your own yarn cake, in a new flavor for every fleece you meet.

I often experience a deep connection to the group I'm teaching. We are there together, in the same room, with the same fleece, doing the same exercises. Each person's skill level and capacity is unique, but still, the connection between us ties us together, in all our struggles, explorations, and successes. We take parts of each other's journeys and create our collective journey. As you read this book, think of it as that classroom and know that all the other readers share your struggles and successes. Know that we have your back just as you have ours. Sense the connection to spinners before, beside, and after you.

Robin Wall Kimmerer writes in her book *Braiding Sweetgrass* about indigenous knowledge and how she learned that the only way to understand something fully is through all four aspects of our being—mind, body, emotion, and spirit. It is my intention in this why-to guide for joyful spinning to illustrate wool preparation and spinning in cognitive, physical, sensory, and spiritual perspectives. This is how I listen to the wool. I invite you to listen with me. Some of these aspects may strike a stronger tune for you than others. That is okay. Perhaps you will be surprised to see that the other whys complement your strongest why. The wool is your most important teacher. I'm just your guide to finding your way to listening to it.

WAYS TO READ THIS BOOK

You can read this book in several different ways. Perhaps you haven't prepared wool from fleece before and want to learn. Or you may be an experienced spinner who wants to dive deeper into spinning and wool preparation. You can use this book as a workbook and create a case study with one fleece through the whole book. Perhaps you work together with a guild or a spinning group.

This can also be a handbook to consult when you stumble upon a challenge. Reread the chapter on picking the fleece when you forget the benefits of it. Dive into the chapter on the knowledge of the hands to remind yourself to trust them. Or use the book as a friend to chat with when you have a setback or a period of low inspiration. *Listen to the Wool* can also be a companion when you explore your own processes or expand your horizons. If you are a beginner, this book can be a companion to a spinning course or a more technical book.

There is an exercise at the end of each chapter. You are welcome to work on these exercises to any extent you like. In most of them I encourage you to explore some aspect of spinning and make notes of what you find. Feel free to record your thoughts in any way that fits you: through notes, drawings, or audio. To your aid I have also provided a glossary and two appendixes, one with references to images throughout the book and one about the tools I use and what I look for when I buy tools.

In the five parts of this book I share my experiences as a spinner and spinning teacher and guide you through different aspects of wool preparation and spinning. In part I, "Explore," we look at the wool on the sheep, the smorgasbord of breeds we have locally and what fiber types we can find in a fleece. I guide you through how I wash my fleece and then pick it before I move on to storing or preparing it. Through all these steps we have the perfect opportunity to explore the fleece and prepare not only the wool but ourselves for the upcoming steps.

Part II, "Play," takes you through preparing the wool—teasing, combing, and carding. These are small and seemingly time-consuming steps, but steps that teach us about how the wool works. This is where you become the expert of the fleece you have in front of you. We also look at different ways to keep records of the wool and what we do with our notes and samples. By playing with opposing ways to prepare the wool, we can land in a way that works with the wool and the tools we have, and match it with our skill level and experience.

Part III, "Create," is where you find the spinning process and how we can find the ease, curiosity, and joy in both spinning and preparation. We look at the Twist Model as a tool for understanding how wool works in states of spun and unspun, and in the spectrum between these stages. We explore woolen and worsted spinning and how we can play with these techniques using different tools. In this section we also look at how we can be kind to ourselves when we spin. Finally, we look at different ways to be in the spinning experience.

In part IV, "Expand," we stretch the spinning process and dip our toes into aspects that can increase the experience as we spin. We look at the superpowers of slowness and of working with simple tools close to our bodies to understand the mechanics of spinning—in the tools as well as in our bodies. We explore the idea of talking and how expressing our experiences can be a powerful tool in understanding the spinning process. We will also look at our mistakes from a perspective of humility toward what we have learned through making them.

Finally, in part V, "Listen," we go deep into our hands and our minds and the connection between them, an important perspective in all crafting. We switch hands and challenge our comfort zone, thereby learning more about what the hands really do and how we can ease the strain on our bodies. We go inward and look at some of the processes that roll as we spin—rhythm, dance, memory, and creativity. All of these can add to the experience when we spin. We land softly in listening to the wool as our most important teacher.

A Note on Breeds

I work with breeds that are close to me where I live in Sweden, and these are the breeds I write about in my examples. You may never get to experience wool from Swedish breeds, but you don't have to. I invite you to learn to explore the breeds you have close to you and to translate my methods of exploring fleece from my home breeds to the breeds that are accessible to you. Some may work, and some may not, but do your best to do your local breeds justice. This is not about right or wrong; it's about the process and finding ways to work from fleece to yarn and textile with the tools, skills, context, and breeds you have.

Compass

As my guide when I make the journey through a fleece, I bring my compass. It has five signposts to ease my navigation and find the way that feels right for me at that point and with that fleece. I am lending you my compass for you to navigate through your journey.

Go Back—to Go Forward

Whenever I struggle with something in my spinning, be it a technique, a decision, or the flow, I take a step back. Back to what I know. Picture yourself in the woods. You turn west at the big rock and realize this isn't the way. You go back to the rock. Now you don't only know the way to that rock in the first place, you also have information about what is west of the rock. By going back, you will still be going forward, but with the new knowledge you gained from that detour. With that in your backpack, you will be able to make more informed decisions for your journey ahead.

I will also encourage you to go back in another dimension. Whenever you feel lost, go back to the sheep that gave you the wool. What did the wool do for the sheep? How can I translate the main characteristics of this wool into a yarn? How can I honor the sheep in the yarn I make? Ask yourself these questions and let the response guide you further along on your journey.

Go Easy—to Go Far

Spinning and handling wool is a joy. To me and, since you have this book in your hands, perhaps to you as well. And it should stay a joy. I don't want spinning to cause strain in my body or harm to the wool. Listen to your body. If it hurts, adjust. Be it another chair, another tool, lifting feet, changing arm positions, or just taking a break. Make sure the spinning stays a joy.

Go Wild—to Go Wide

I don't spin art yarns. I don't find them particularly interesting or useful. The few times I have spun art yarn have been at spinning competitions, where someone else has stated rules for my participation. And I have loved every competition. Having someone else decide what I should spin—still with room for my artistic expression—has taught me a lot. Sometimes I don't feel confident at all in these techniques, but I go through with it anyway. The worst thing that could happen is that I learn something new. Perhaps I can use my new experiences for yarns that are closer to my comfort zone. Perhaps I even like the new technique and welcome it.

Another way to go wild is to break the rules you have learned about spinning or wool preparation. Reading or learning about a rule, whatever it may be, is one thing, but to really understand it I need to understand it in my body and in my own experience. So sometimes I knowingly break an established rule, just to find out what it means. It's like a rule of physics. I know the rule, I have the condensed expression in my mind's eye, and I trust all the scientists who have developed it; however, by challenging it I will learn why the rule looks the way it does, how I got to the same conclusion, and, perhaps, when it is possible or even advisable to break it.

Remind yourself every now and then to get outside your box, outside your comfort zone, or at least stretch it a bit. Go wild.

Go Slow—to Go Deep

I will talk a lot about superpowers in this book. One of the most important superpowers for me is the power of *slow*. When I go slowly, I get to understand the process, the technique, and the characteristics of the wool. By going slowly—with a spindle instead of a spinning wheel, for example—my hands get time to feel the wool, how it behaves in the spinning, and how my body and the spindle work together in the spinning mechanics. Going slowly gives me time to understand on a deeper level—cognitively, sensorially, physically, and spiritually.

Go Wrong—to Find the Right Way

It's easy to land in some sort of default wool processing or spinning technique. But there is so much more out there! To find a way through a fleece I encourage my students to go wrong, to try techniques that they don't believe in for the fleece in question. If you see a fleece as a perfect candidate for woolen spinning, try spinning it worsted. If you are convinced that a fleece will look its best separated into undercoat and outercoat, try processing it all together. You may come to one or more conclusions: you were right about your original idea—it worked just as you had anticipated. Good, then you have more substance to back up your original plan. Or you were wrong. The technique you didn't believe in worked excellently and made you change directions. Good! You have now learned something new. Trying different techniques, even the ones you think won't be the best for your fleece, will help you deepen your understanding of your fleece by sharpening the tools you have in your toolbox or by adding new ones.

A Box in My Lap

September 2011. I am sitting in a circle of women in a barn at a city farm. It's the first day of my first spinning course. I have a box in my lap, filled to the brim with wool. On the side the name *Pia-Lotta* is written, above "Finull, 900 g." Pia-Lotta is the Finull lamb that was just freed of her summer coat. She is skipping around in the pasture right outside the barn, and I am responsible for her fleece.

A couple of weeks earlier I had been talking to my knitting friend Anna. She told me about a shepherdess friend who threw away the wool from her sheep since she didn't know what to do with it. It turned out that, at the time, more than 90 percent of Swedish wool was thrown away, buried, or burned. All the while we were importing tons of wool from New Zealand. Then and there Anna and I decided to learn

Figure I.1. Raw fleece from Hanna the Gestrike sheep.

how to spin, to be able to take care of a tiny part of Swedish wool that was otherwise going to waste.

The barn smells of farm and sheep. The wool in the box has slim and crimpy staples, creamy white. When I examine them, I see the fibers that make out their shape. They have decided to meander along the same waves together, from cut end to tip end. If I pick up a handful of staples and let go, they scatter down into the box one by one, like snowflakes. I inhale a fresh, moist smell. Later I find out that it is the grease of the wool, the lanolin, that wraps my mind in kindness.

My hands wade their way through the wool mass, finding signs of Pia-Lotta's life in the summer pastures. An occasional piece of hay, a juniper needle, tiny seeds, and other treats from the ground she calls home. Deeply buried in the fleece I find a perfectly round horse chestnut.

"Now we will tease the wool!" says the teacher, Marie. I have no idea what that means; my mind has been busy picking the staples. I pick up one and hold it vertically. With my other hand, I slowly pull a few fibers sideways, as instructed. I watch the fibers move in a bow from the staple, linger for a while, and eventually give in to the intruding air and let go. Staple by staple transforms between my hands into an airy wool mass. Pia-Lotta wouldn't recognize her own coat.

After having been mesmerized by the transformation of the fibers, we card the wool. I don't receive many instructions about how to card, but I do card the teased wool and end up with rectangular arrangements. I spin them on a large and heavy spindle from the barn shop. I can't remember how we were instructed, but yarn did come out of the spindle. Uneven, overspun, and gloriously beginner-y—my first skein of handspun yarn.

The spinning course laid the foundation of my perspective on spinning. The fleece the sheep grew from the grass she was grazing is the start of my spinning process. By letting in air between the fibers, step by step, I learned—and still learn—to understand how a fiber works and how I can work with it. I have access to previous information in each subsequent process and can use it to understand the following and make further decisions from that. Through the gradual disintegration of the staple, I get the time and space to explore the wool.

CHAPTER 1

Off the Hoof: Listen to the Sheep

11,000 years ago. Domestication of wild sheep is believed to have begun in Mesopotamia, perhaps also in other locations. The wild sheep has large horns, a short tail, and a hairy and dark coat that molts annually.

8,000 years ago. Evidence shows sheep in Anatolia, parts of western Asia, Cyprus, Crete, and southeastern Europe.

7,000 years ago. Sheep appear in northern Africa.

6,000 years ago. Sheep reach the British Isles and China.

5,000 years ago. The biggest changes in sheep domestication have led to smaller horns, a shorter tail, and a white fleece. Sheep are hairy or woolly, but less hairy than their ancestors and look similar to today's Soay sheep (from the island of Soay in the Scottish archipelago of St. Kilda). Sheep molt their wool once a year, resulting in a yearly wool yield of 10–30 ounces (300–900 grams). Because of the relatively low wool yield, large flocks of sheep are necessary for wool production. Spindle whorls of 0.3 ounces (8 grams) are used in Greece in this time period. The low whorl weight indicates deep knowledge of spinning, with the argument that preparing for and spinning fine yarn takes more skill than spinning thicker yarn.

4,500 years ago. In the mosaic Peace Panel in the Standard of Ur, sheep that can be interpreted as both hairy fat-tailed and wooly short-tailed are depicted.

4,100 years ago. Tablets in Ur in Mesopotamia show a grading system with three categories: property of the Moon God, Royal, and Mixed.

4,000 years ago. On other Ur tablets, calculations are made of the time and work force needed for the manufacturing of 9 pounds (4 kilograms) of raw wool into 11 x 11 feet (3.5 x 3.5 meters) of woven tabby. Washing and preparing the wool will take one person thirty-seven days, spinning the wool will take one person sixty-seven days, setting up the loom will take three people three days, and weaving the fabric will take two people seven days.

3,800 years ago. Personal communications—more than twenty-three thousand tablets—reveal how much wool and textiles the merchants in Anatolia need their partners (usually wives) in Assur to send for the markets.

3,500 years ago. More than forty-four hundred spindle whorls are in use in a one-acre settlement with around 120 inhabitants in the Po valley in northern Italy.

3,300 years ago. According to tablets, the palace in Knossos, Crete, owns up to one hundred thousand sheep. The tablets also estimate the annual wool yield from one sheep at an equivalent of 26 ounces (750 grams). Bodies buried in oak coffins in Denmark wear wool clothes. Recent archaeological techniques show that the wool in the clothes is imported.

3,200 years ago. Textile finds in the Hallstatt salt mines in Austria show a wide variety of weaving techniques and dyes.

2,600 years ago. A Babylonian glossary tablet lists words for both plucking and shearing sheep.

29 BCE. Roman poet Virgil writes in the agricultural poem "Georgics" about selective sheep breeding for the benefit of wool.

Eighth century CE. Villages in half of the counties in England have sheep or sheep-related words in their names.

Ninth century CE. Vikings bring short-tailed sheep to Iceland. In the beginning of the tenth century, initial measures are taken to prevent further importing of sheep to Iceland.

Twelfth and thirteenth centuries. The Spanish royalty import rams from the Beni-Merines in

Figure 1.1. Jämtland (white and one brown), Gotland (black head and legs), and Gute (horned) ram lambs. Notice the difference in size—the Jämtlands (with some Merino and Texel in them) are larger than the Swedish landraces Gotland and Gute sheep.

what today is Morocco and cross them with their best sheep. The new breed is later given the name Merino.

Early fourteenth century. Wool producers in Yorkshire export nearly seven hundred sacks of wool annually to Italian buyers.

Fourteenth century. Several paintings around Europe depict polled sheep with mainly short wool, white faces, and long tails.

Mid-fourteenth century. A man is killed in a bog in Bocksten outside of Varberg in Sweden. Through the acidity of the bog, his woolen clothing has been preserved. Analyses show that his cloak had been spun of wool that resembles the wool from Swedish Gute sheep (see chapter 2), with a slightly higher proportion of rough outercoat fibers.

Sixteenth century. Spanish colonists bring Churro and Merino sheep to the Americas.

Nineteenth century. British colonists bring Merino sheep first to Australia and later to New Zealand / Aotearoa.

1985. Icelandic sheep are exported to North America.

Wool has developed into a material that grows out every year and has many unique capacities:

- Wool has high insulation against both heat and cold. This happens through the crimp in the fibers that keeps them apart and invites air.
- Wool actively generates heat by absorbing moisture vapor from the air into its interior structure. This is known as heat of wetting.
- Even when wet, wool will keep the wearer warm. By enticing water on the surface of each fiber to split into separate drops, it prevents heat energy from traveling along them or hopping across to neighboring fibers, retaining its insulating capacities.
- This same repellent capacity of wool fibers means that a stain usually stays on the surface of the textile and can easily be removed.
- Wool fibers can absorb up to roughly 30 percent of their weight in moisture vapor through their porous surfaces to their moisture-loving internal structures.
- Wool is naturally fire resistant through this capacity to hold moisture.
- Odors from body sweat are captured by these interior structures for days or even weeks, leaving wool clothes almost odor free during use. They self-clean the next time the fibers are aired in damp air. Therefore, a wool garment doesn't need to be washed as frequently as garments in other materials.

The properties of wool and how much we owe it through the millennia make me gasp for air. I want to give back, make the wool as beautiful as I possibly can, and humbly share my experience with the world. Sometimes, though, I stumble, wondering: Who am I to spin this magnificent fiber? Then again, in my next breath I ask myself, with eleven thousand years of breeding sheep for their wool, who are we to burn it today?

Responsibility

Isn't it our responsibility toward the sheep, the sheep farmers, and the earth to recognize the value of wool and make good use of it? Reasons for burning, burying, or otherwise wasting wool can be that the sheep owner doesn't know how to take care of the wool, because they don't know how or where to sell it, because of low demand or because it is more expensive and time consuming to take care of it than to waste it. Since my first spinning lesson in 2011 (see "A Box in My Lap" on page 1), the waste has decreased—in 2016 to 73 percent and in 2020 even further, to 54 percent. Wool, however, is still classified by the Swedish Board of Agriculture (Jordbruksverket) as a byproduct of the meat industry and not as a product with its own value.

Many initiatives, large and small, have been taken to reduce the wool waste in Sweden.

- The Gotland company Ullkontoret (the Wool Office) is the only scouring mill in Sweden. It started in 2016, sparked by the high amount of wasted wool in Sweden. Gotland has the most sheep in Sweden, and the founders of Ullkontoret realized that, to take care of any wool, it needs washing first, preferably within the borders of the country. Today the mill scours 2 tons of wool daily during spring and autumn, from both the island and the mainland. In 2024, Ullkontoret started their own spinning mill.
- Ullförmedlingen, the Wool Broker, started as a Facebook group in 2017 and later launched its own website. The broker strives to connect sheep

owners and wool buyers, whether they are hand spinners buying one fleece or an industry looking for 500 kilograms of wool, to decrease the high amount of wasted wool.

- Arena svensk ull was founded in 2022, with an aim to strengthen the cooperation for Swedish wool and to increase its use. The organization gathers lamb producers, shearers, wool industry workers, wool stations, spinning mills, wool manufacturers, crafting organizations, wool counseling, and educators.
- The first Swedish wool classification system was established in 2023 (see page 9).

Similar stories can be seen in countries like Estonia, the United Kingdom, United States, and Canada—tons of wool being wasted, but also initiatives and solutions for reducing the waste.

Fiber Types: What Is the Fleece to the Sheep?

As we have seen, sheep have been bred for their wool through millennia. It's just relatively recently that there has been a decline in the use of wool, in favor of first cotton and later petroleum-based materials. But the wool is still there, growing out every year, and it is a natural product with outstanding capacities. To find out what the wool can do for us, we need to look at what the wool is to the sheep and how it works to protect the sheep against the elements.

Sheep's fleeces are built up of hair fibers and wool fibers. Hair fibers are straight, smooth, long, and strong, and we refer to them as outercoat fibers. They grow out of primary follicles that start developing around sixty days after conception. Through the usually coarse primary follicles the hair fibers grow fast, coarse, and long.

Wool fibers, on the other hand, are fine, often more crimpy and more elastic. They grow out of the secondary follicles that arise around 120 days after conception and keep developing to about 30 days after the lamb has been born. The growth of the wool fibers in the secondary (and finer) follicles is slower, and the fibers are finer. For this reason, these wool fibers, which we call undercoat, are shorter than the outercoat fibers that cover them. In some breeds, like Merino, the secondary follicles can branch under the skin, resulting in more fibers growing out of one and the same follicle. In breeds that grow consistent

Figure 1.2. Light Gute lamb's wool with black kemp fibers and dark Brännö ewe's wool with white kemp fibers. The two staples in the center are intact, and the ones one step out are teased. The edge tufts is what I have teased out, mainly kemp.

Figure 1.3. Different types of fibers and fiber fineness in staples of Swedish Poll Merino (ram lamb), Gute (ram lamb), and Gotland (ewe) wool.

fleeces, the difference between fibers from the primary and secondary follicles is too small to perceive, as described below. Breeds that have differing outercoat and undercoat fibers are referred to as having dual coats, see also figure 4.5.

Some fibers have a core, the medulla. The coarsest hair fibers with a large and spongy medulla are called kemp fibers. These are usually white, black, or red, and quirky. Due to the large medulla, they are brittle and usually break, and thus tend to shed (figure 1.2 and also samples 6 and 8 in figure 2.1). Kemp fibers have often been bred away in modern breeds, while they can be present in more primitive breeds (read more about kemp in chapter 2).

The fiber types—undercoat, outercoat, and in some cases kemp—work together in staples to protect the sheep. A staple is a cluster of adjacent fibers that are joined by a combination of characteristics, like length, crimp, and speed of growth. The undercoat grows closest to the body of the sheep and keeps the sheep warm. The coarser outercoat fibers reinforce the staple and help keep the undercoat open, inviting air in between the fibers. The outercoat fibers grow out past the ends of the undercoat fibers to form a denser tip. This helps lead water off the body, keeping the sheep dry. If the fleece contains kemp fibers, these also help keep the fleece light and airy, through additional armoring, and in keeping the staples upright for further protection against the elements.

In some breeds, one fiber type has been bred to resemble characteristics of the other fiber type. For example, in a breed where strength and shine are the desired characteristics, there will still be undercoat fibers, but they will be fewer and may have the characteristics of outercoat fibers. The primary follicles have given more space for the secondary follicles to grow and become stronger and coarser. Or, at the other end, in a breed that has been bred for fine fibers, there may still be outercoat fibers, but they may look and feel more like undercoat fibers. The undercoat fibers have been bred to multiply in their follicles so that they take over, making the outercoat fibers lose their advantage in speed and growth. The ratio of secondary to primary follicles can change as well.

Both undercoat and outercoat fibers are covered with scales. In the Swedish landrace and heritage breeds (see chapter 2), the scales are placed like floor tiles, edge to edge. This creates an even structure that reflects the light, making the wool shiny and smooth. This wool tends to felt easily. Breeds like Icelandic, Faroese, and Shetland sheep also have this type of wool. On other breeds, the scales are wrapped over each other, more like roof tiles. The wool looks more matte and doesn't tend to felt as easily. In yet other breeds, the scales are wrapped in different manners.

Other ingredients in the fleece are lanolin and suint. Lanolin is the grease (or wax) that is produced in the skin. The lanolin protects the skin and fiber against wind, rain, and sun, similar to the lipids in our own hair. It also protects the wool from dirt to some extent. Suint is sweat salts. When wool gets wet, the alkaline salts in the suint dissolve and create a natural soap together with the lanolin. This soap, rain, and the sheep's own heat generation, plus the density of the wool and the protective qualities of the lanolin itself, keep the sheep clean.

Primitive Sheep Breed Characteristics

Since the domestication of sheep, we have bred them to suit our needs for wool. Today it is difficult to see the resemblance between the exceptionally short and hairy fleece of wild sheep and some of the Merino breeds with massive- and fine-fibered fleeces. But there are still some breeds that are considered primitive, that have kept characteristics that the earliest sheep had. One such example is the Soay sheep, believed to have roots in the Bronze Age. Soay sheep have horns and are usually dark or light brown. Many primitive breeds have a dual coat that molts, or sheds, usually in the spring. Gradually, the fibers weaken at the end of the past year's growth. This spot is called the rise. The new year's growth comes underneath the old fiber that is about to let go. A shedding sheep can be plucked, or rooed, in late spring or early summer. Where a shedding breed has grazed it is common to find tufts of wool in the vegetation. The tufts, hentilagets in Shetland, have been common to gather.

As the old fleece falls or is plucked off the sheep, there are no cut ends in it. Rather, the "cut" end of a shed or plucked fiber is tapered due to the weakening of the fiber at the rise. The fibers thus have the same cone-shaped ends in both tip ends and shed ends. This has importance for the itch factor. A fiber that touches the skin can either yield to the skin, if it is fine or flexible, or make the skin yield, if it is coarse. The itch occurs when the skin has to yield to the fiber (and of course people have different degrees of tolerance to itch). A cut fiber provokes the skin more than a tapered fiber. This means that a shed fiber that is tapered in both ends can be coarser in diameter than a cut fiber but still generate less itch. I have worked with Shetland and Gute fleeces that have been shorn one-third to two-thirds of an inch (1 or 2 centimeters) after the rise. This means that the staples consist of the tip ends with the past year's growth down to the weak spot, the rise, and then another one-third to two-thirds of an inch (1 or 2 centimeters) of the new year's growth down to the cut ends. In preparing a fleece like this, with the rise included, I have been able to remove the new year's growth below the rise by lightly tugging or teasing (see chapter 7) the staples. This way the fibers in my preparation are tapered on both ends. The rise is usually visible as a waist, or thinner portion, close to the body of the sheep or the cut ends.

Other primitive breeds that can have more or less of the primitive characteristics include Old Norwegian sheep, Icelandic sheep, and Faroese sheep.

Systems for Describing Wool

Systems for describing wool are specific for different countries and depend on the types of breeds and the conditions in the country. Even if the systems are aimed at the wool industry rather than hand spinners, getting an overview of some of them may help us understand how wool can be categorized and what characteristics are valued—and not—on the wool market. As hand spinners we have the opportunity and the advantage to see beyond standards adapted to a market that values homogeneity in the wool. We can embrace and dive into the diversity of smaller breeds and characteristics that a spinning mill can't make use of on a bigger scale. I like to see the standards of the wool industry as a grid that stops additional exploration for the industry but allows me to reach through and explore further. For example, in the Swedish wool standard, the Swedish allmoge breeds—eleven individual heritage and conservation breeds—are bundled into one category together with Rya and Gute sheep. These are all breeds that are seen as too variegated or complicated for the demands on the wool market. As a hand spinner, though, I can dig deeper and find gold (see chapter 2 for descriptions of the sheep breeds in Sweden).

For an excellent overview of a wide selection of breeds and their wool from the hand spinner's perspective, see *The Fleece & Fiber Sourcebook* by Deborah Robson and Carol Ekarius.

Sweden

Traditionally there has been no systematic way to classify wool in Sweden. However, the many initiatives during the past few years for taking care of Swedish wool (see pp. 5–6) have also generated a Swedish wool standard, published in 2023. The system is directed toward players on the industrial wool market. The wool standard sorts wool into categories based on wool characteristics, like length, diameter, color, and character, rather than breeds. The Swedish wool standard is built on the Norwegian wool standard and adapted to Swedish conditions. It was created to give a common ground for players on a wool market that consists of lots of small-scale sheep owners. With the classification system, different breeds with similar characteristics can be bundled together for larger batches of wool. The standard includes recommendations for sheep farmers and shearers, plus the actual classification system, with explanations of the different parameters and methods of analysis. There are six quality types, from a micron count of under 21 at the top to over 35. The main categories have subcategories describing other characteristics like color, degree of vegetable matter, fiber length, and general character. Another three quality types describe bouncy wool like that from typical meat breeds, highly variegated wool like the Swedish allmoge breeds (see chapter 2), and wool that is sorted out, like belly and crutch wool.

Norway

Norwegian wool is sorted into one of three categories—Crossbred, Spælesau, and Pelssau wool. NKS (Norsk kvit sau, Norwegian white sheep) is the most widespread of the Norwegian breeds and is a crossbred with white and even fibers with a high bounce. NKS lacks the long and strong outercoat fibers that many of the Norwegian breeds have and consists of mainly undercoat fibers. Spælsau breeds belong to the short-tailed group (*spæl*, meaning "short tail"; *sau*, meaning "sheep") and are the oldest breeds in Norway. They are dual coated with a long, strong, and shiny outercoat and a short and fine undercoat. Pelssau sheep are originally a cross between Swedish Gotland sheep (see chapter 2) and blueish gray Spælsau. The breeding has been focused on the quality of the skins.

The Norwegian wool standard is based on these three groups of breeds, with the finest crossbred wool at the top and divided into six quality groups with subcategories based on factors like what season it has been shorn, degree of vegetable matter, cotting, length, color, crimp, and character.

United Kingdom

The purpose of the British Wool Marketing Board is to work as a link between sheep farmers and the textile industries and provide an organized structure to manage supply and demand of British wool for an international textile industry. The more than sixty breeds in the United Kingdom are generally adapted to the part of the country in which they graze, as the weather and geographical conditions influence the wool characteristics. The board has organized British wool into seven categories: fine, medium, cross, luster, hill, mountain, and naturally colored. Each category has several breeds belonging to it, together with a micron range, typical staple, and main end uses. The categories are further described regarding general wool characteristics and those of individual breeds within the category. The UK system is thus a combination of a classification system and a description of typical wool characteristics for the categories and the individual breeds within them.

United States

In the United States, four different systems have been used for classifying wool:

- Bradford Count (originating in the United Kingdom) focuses on the number of hanks (560 yards) of yarn that can theoretically be spun from a pound of wool. A high number in the count system is a fine wool, and a low number is a coarser wool.
- The Blood System is based on the breeding of the sheep, divided into six grades from fine (Rambouillet or Merino) at the top and half blood down to one-quarter blood, Common, and Coarse. The system originated from the early colonization and can't describe newer breeds.
- Spinning Count measures the fiber diameter in micrometers, or microns, from superfine (up to 19 microns) to coarse (from 33 microns). The system

includes sixteen grades and requires a lot of training to use. The system is no longer a standard.

- USDA (United States Department of Agriculture) Grade uses the same numbers as the Spinning Count system but has an average micron count interval assigned to each number, plus standard deviation. The system requires lab analysis. This is the primary system currently used in the United States.

The Bradford Count, Blood System, and Spinning Count are subjective, and the USDA Grade is objective.

Australia

In Australia, sheep breeds are divided into two categories: Merino and Crossbred, where Crossbred refers to non-Merino breeds and crosses. Six breed groups with designated micron ranges from low to high are used in appraising and typing wool. From the breed group codes, additional codes are used to specify wool category, style, type of vegetable matter, and characteristics like fiber length and strength, plus standardized comments.

EXPLORE: GREET

If you have a fleece, have a look at it and see what you find. Can you see staples of different lengths? What does the lanolin feel like? Can you identify different fiber types in one staple? How are they alike and different? If you are working with a primitive breed like Old Norwegian, Shetland, Icelandic, Faroese, or Gute sheep, can you identify a rise? If you have a microscope or a loupe, have a look at some fibers through them and see what you find. Greet your fleece and take notes. You don't have to know the established words, just jot down your observations, reflections, and questions.

If you don't have a fleece yourself, perhaps you can look at a friend's fleece, or come back to this exercise when you have one of your own (see tips on finding a fleece in chapter 2).

CHAPTER 2

A Smorgasbord of Wool

I love wool in all its shapes and forms, but my favorite phase of wool is its raw state, just shorn off the sheep. In my mind I can hear my own squeak of joy as I dig my hands into a newly shorn fleece, sensing the rich lanolin in my hands.

My husband Dan and I take quite a few trips around the countryside to visit sheep farming friends for a chat and a photo shoot with their flock. One such trip was to Claudia. She has a fourteen-head flock of Värmland and Gestrike sheep (see pp. 18–19 and p. 173).

When we get to the pasture and are welcomed by a group of brown sheep, Claudia tells us that the behavior of her two breeds is remarkably different—while her brown Värmland sheep are instantly curious and up for adventures, the gray Gestrike ladies are more reserved, at least in the beginning. Both breeds are quite cuddly, though, and after a gray-headed conference in the periphery of the pasture, I am surrounded by both brown and gray sheep, all nudging me for attention.

I dig my hands into the thick fleece of a Gestrike sheep, Härvor. I feel her warmth spread through the staples into my hands. The lanolin glistens in the October sun; the smell of sheep slows my heart down into a calm rhythm. Härvor's head is close to mine, and I feel her exhales warm against my cheek. We exchange breaths, hers become my inhale and mine become hers. Her fleece encloses my hands, the staples she has grown since the spring shearing. The nutrients in the pasture have fed her through the summer, helping her grow a fleece to keep her safe.

A few days later I come back for shearing day. Elin the shearer works through all the sheep in around thirty-five minutes. My job is to quickly survey the newly shorn bundle that Claudia brings to me, together with a paper bag with the sheep's name, all during the few minutes Elin shears the next sheep. I spread the treasure onto a grid and start picking out whatever vegetable matter, second cuts (not that there are any), felted parts, and poo I can find. As soon as the buzz from the shearing machine stops, it's time to put the fleece in its bag and be ready for the next one, together with a bag with another name and the date of the shearing hastily scribbled onto it. Some of the bags have crossed-over names and older dates.

As I receive each newly shorn fleece, I can still sense the warmth of the sheep that just wore it; her breath is still present in the mass of moving staples in my arms. I drink the smell of fresh lanolin and see the glittering and abruptly cut ends that just a moment ago were part of a moving body. Any vegetable matter I find are remnants of the sheep's own living room, pieces of her life that fall out of the shorn fleece as souvenirs from her home.

It dawns on me that I wouldn't be a spinner without sheep like Härvor or sheep farmers like Claudia. The fleece Härvor grows and gifts to me, and the love, dedication, and hard work Claudia puts into her sheep and pastures make me a better spinner. How could I then not spin this wool from my heart? How could I then not create the best yarn Härvor's fleece can be? How could I then not return the gift by sharing my words with the world? Härvor, now bald and slim, grunts quietly in the crook of my arm. In my mind I lean in and hear her whisper softly, "Listen to the wool." And I will. There is my answer, there is my connection—back to the pastures, back to the soft breath and whisper of the wool.

It's in the raw fleece that my spinning journey begins, in the coat that the sheep has grown and now been relieved of. It is in the characteristics of the unique fleece that I can dive in, explore, and create a coat of my own. To do that I need to understand the characteristics of the wool and what they do for the sheep.

As spinners we may have access to breeds around the globe, but the closer the sheep are to us, the easier it may be to feel a connection to them. We have

Figure 2.1. Samples of wool from nine individuals of the same flock of Brännö sheep.

the opportunity to explore different kinds of fleece, qualities, and wool types closer to us than we may think. In this chapter, I want you to discover and explore the wool you have locally. I give you a presentation of the wool types and breeds that are local to me but, more important, the characteristics any wool can have. There will also be a section where I share my thoughts on buying a fleece, what I am looking for, what I stay away from, and what melts my heart.

Swedish Wool Types

Traditionally, Swedish wool has been described through four categories of wool, or staple, types. These are characterized by the shape of the staple and largely depend on the ratio of undercoat to outercoat in the staple.

A fleece from a Swedish breed can consist of one wool type only or be a mix of two to four wool types, all originating in the Swedish landrace (figures 2.1 and 2.2). A combination of wool types is common in the Swedish allmoge breeds (see below). Some breeds have breeding goals aiming toward a specific wool type, while others can have variegated fleeces with all the wool types.

Rya Type Wool

Rya-type wool (not to be confused with the Rya breed or the rya textile made of rya knots) usually has 40:60

Figure 2.2. All these staples come from the autumn shearing of Härvor the Gestrike sheep. Her fleece includes examples of all the staple types of Swedish landraces. Leftmost are finull-type staples, next are vadmal type, the shiny wavy ones are gobeläng type, and on the far right are rya-type staples.

ratio of outercoat and undercoat, respectively. The staples are generally long and conical, with an airier base and a denser tip end. There is rarely any crimp in rya-type staples. The combination of long and strong outercoat fibers and soft and airy undercoat fibers makes this wool type versatile. The hand spinner can choose to use the fiber types together, separated, or semi-separated (see the section on Rya sheep below for more background of the rya-type wool). The wool of Rya sheep is aimed at rya-type staples, but many of the allmoge breeds (also below) can have rya-type staples.

Finull-Type Wool

Finull-type staples look as if they have mainly fine and quite short undercoat fibers, but in fact they have very few undercoat fibers and consist mostly of outercoat fibers that are unusually fine and short with fine crimp in well-defined staples. Wool of finull type is mainly used to spin soft and airy yarns. Finull sheep (see below) are bred for finull-type wool, but many of the allmoge breeds can also have this wool type (figures 2.15, 4.2, and 7.2).

Vadmal-Type Wool

Most of the fibers in vadmal-type wool are undercoat fibers, but a few outercoat fibers peak up above the undercoat base, creating a wide-based triangular staple of a length between rya- and finull-type staples. The name *vadmal* means "a heavily fulled woven textile," and the term in this context refers to the kind of staple that would work well in that kind of fabric, with the undercoat fibers responsible for most of the felting and the outercoat fibers armoring the structure with their length and strength. Vadmal-type wool can be found in many of the allmoge breeds, although usually quite sparsely (figure 2.14).

Gobeläng-Type Wool

The gobeläng-type wool (also called päls-type wool) is strong, shiny, and free from kemp. It consists of mainly outercoat fibers and has well-defined waves or locks rather than crimp. Gotland and Swedish Leicester sheep (below and figures 2.5 and 8.13) have breeding goals aiming toward päls ("pelt") or gobeläng wool, but some of the allmoge breeds can also have this staple type. The word *gobeläng* translates to "tapestry" and refers to a strong and lustrous type that would suit as a yarn for Scandinavian-type tapestries.

Swedish Sheep Breeds

All the Swedish landraces and heritage breeds are of the Northern short-tail kind, quite small and with wool-less legs and faces. The fleece is usually quite low in lanolin. Five of the Swedish breeds have explicit breeding goals for the wool, three of which emphasize the wool for the sake of the wool itself, the other two for the skins.

In the descriptions of the landrace, heritage, and other breeds that follow, I focus on the history of the breeds in Sweden in general and the wool for spinning purposes in particular.

Gute Sheep

The most primitive sheep in Sweden is the Gute sheep, stemming from the Gotland outdoor sheep, Gotländskt utegångsfår. The breed has many characteristics that are significant for a primitive breed—horns in both rams and ewes, soft undercoat, strong

Figure 2.3. Stina is a cross between Rya and Finull sheep. Just like all the Swedish landraces, heritage, and conservation breeds, she is quite small and has a wool-less face and legs.

Figure 2.4. A Gute ram lamb.

and coarse outercoat, kemp, and mane hair. The wool comes in many colors from white to black but occasionally also brown. In the spring, the wool molts and can be plucked (see chapter 1). The presence of kemp (figures 1.2, 2.1, and 2.17) keeps the fleece light and airy, protecting the sheep from cold or heat, and reduces the need for crimp. Both the look of the sheep and the characteristics of the wool correspond to many archaeological finds.

In the 1920s, a program was started to breed polled (hornless) sheep with even and shiny wool, suitable for skins, using Gute sheep. The program later resulted in the Gotland sheep. A few sheep of the Gotland outdoor sheep were saved, though, and these were the foundation of the Gute sheep population we have today. A Gute ram is the symbol on the flag of the island of Gotland.

A Gute fleece is a fleece of great versatility and contrast and, in my experience, a perfect beginner's fleece. In a single fleece, I can find outercoat fibers that are long and strong, sometimes coarse, sometimes less so. The undercoat can be anything from medium to extremely fine, especially in a lamb's fleece. Due to the presence of kemp, the staples are light and airy and easy to work with.

Gotland Sheep

The story of Gotland sheep (see the gray sheep with black head and legs in the center of figure 1.1) begins with the story of the Gute sheep above. The first name was actually Pälsfår ("fur/pelt sheep," to refer to the beautiful skins), but later changed to Gotland sheep to accommodate an international market. And it has become an international market—flocks of Gotland sheep have been exported to the United States, United Kingdom, Australia, and New Zealand. If you

Figure 2.5. Raw fleece from Gotland sheep.

have experience of Gotland sheep outside of Sweden it is likely that their fleeces don't look and feel like the fleeces of Gotland sheep in Sweden. The breeding goals for the wool aim toward even and three-dimensional locks over the body of the sheep; an even color of light gray, dark gray, or black; and a silky shine. These aims for the wool have been constructed for the sake of the skins rather than for the shorn wool. The fibers are mainly outercoat or outercoat-like undercoat fibers and typically form päls-type staples. The micron count is aimed at 35–45.

Since Gotland sheep is the most common breed in Sweden, Gotland wool is also common in Swedish spinning mills. But because the wool is so slippery and long, it doesn't always work well with the spinning machines. Quite often the Gotland wool is mixed with up to 30 percent Finull wool (below). Spinning Gotland wool by hand can be equally challenging with the slippery fibers.

Rya Sheep

Rya sheep is the breed that looks the most like the old Swedish landrace. Rya sheep have a particularly interesting history. The name *Rya* originally referred to a rya textile, a woven textile with rya knots, creating a pelt-like structure. These were used as bed covers, pelt-like side down. An inventory list for the nuns in Vadstena convent from the early fifteenth century, for example, states that the nuns should have a rya rug to sleep under during the summer. During the national Romantic era at the turn of the last century, when old textiles were examined, a difference was detected between the older rya rugs and the more modern, postindustrial samples. During the Industrial Revolution, the long wool from the Swedish landrace was not desirable, since it didn't work well in the spinning mills. Therefore, more modern breeds were imported, and the Swedish landrace was suppressed. Arts and crafts managers realized that there was a possibility that the sheep with the shiny wool that had been found in the older rugs were still out there. And they were. Three flocks of what is today known as Dalapäls sheep (below) were used to form a new breed that got the name Rya sheep, whose wool resembled the wool in the preindustrial rya textiles.

Today the breed standards for Rya sheep have a wool assessment system with the aim of using the

Figure 2.6. Rya sheep.

Figure 2.7. A black Finull sheep (with bleached tips) in the company of Texel sheep.

wool in textile crafts. Wool from Rya sheep can be white, brown, gray, or black. The breed standards aim to keep the colors separated. Especially the white sheep should be free from pigmented fibers. The staples should have a maximum of two waves per 2 inches (5 centimeters) and should be at least 6 inches (15 centimeters) long at the age of 120 days. The staples should be even in length and quality over the body of the sheep. Rya sheep wool is a dual coat, usually with 50 percent strong, shiny, and long outercoat and 50 percent soft, fine undercoat (compare to the description of rya-type staples above). A fleece from a Rya sheep can therefore offer many possibilities, depending on how the spinner uses the different coats–together, separated, or semi-separated. Commercially, yarn from Rya wool is popular as a tapestry yarn.

Finull Sheep

Just like the Rya sheep, Finull sheep stem from the old Swedish landrace sheep. In the sixteenth century, the king wanted a finer wool than the landrace sheep could offer. He decided to import sheep with finer wool and aimed to extinguish what he considered to be the horrible Swedish sheep. Luckily, Swedish farmers needed the landrace for a wide range of uses and kept their flocks. After World War I, some enthusiasts showed an interest in the Swedish landrace and found reasonably unmixed flocks in remote areas in Sweden. Until the 1930s only one landrace was registered, but in 1936 three wool types were distinguished: päls (fur or pelt), rya, and finull (fine wool). A few attempts were made to cross the fine-fibered landrace sheep with the Finnish landrace, and through this there is a small degree of kinship between the Finnish landrace and Swedish Finull sheep, more so in the brown variety. The breed association for Swedish Finull sheep was formed in 1995, distinguishing them from the Rya sheep.

The wool from Swedish Finull sheep is white, black, or brown, fine and with a high crimp. The wool is one of the emphasized characteristics in the breeding goals, aiming for a soft, shiny, and fine-fibered wool, 20–30 microns in fiber diameter, and even quality over the body of the sheep. As with the finull-type wool (see above), staples of wool from

Finull sheep look like they mainly consist of undercoat fibers but do, in fact, consist of mostly outercoat fibers that are unusually fine and short. Due to the fineness of the fibers, wool from Finull sheep is popular among both hand spinners and spinning mills.

Swedish Heritage and Conservation Breeds: Allmogefår

Starting in the sixteenth century and more intensively during the first part of the twentieth century, the landrace sheep were replaced with more productive breeds (from a meat perspective) with a higher slaughter weight. The eleven breeds we today call allmoge breeds stem from flocks of the Swedish landrace that have been isolated in remote areas for a long time, resulting in genetically unique breeds. The word *allmoge* literally means "peasant or countryside people," although an allmoge breed is a protected cultural heritage breed. All eleven allmoge breeds have gene banks, making them both heritage and conservation breeds. They are all named after the geographical place where they used to be isolated. Most of the gene banks started in the 1990s.

Due to the small number of individuals in the allmoge breeds, sometimes less than forty breeding ewes (according to the official statistics from the Swedish sheep breeding association from 2023), the breeding goals include keeping the genetic variation in each breed. Therefore, it is not advised to breed for or against specific characteristics, like the wool. The wool from the allmoge breeds can as a result be variegated between flocks, within a flock, and even over the body of an individual sheep (figures 2.1 and 2.2). To a hand spinner, a fleece with different wool types can be interesting. A friend of mine who shepherds two allmoge breeds calls these types of sheep "household sheep"—the fleece from one individual can be used for a wide variety of yarns and purposes in a household.

Värmland Sheep

The Värmland sheep is the largest of the allmoge breeds when it comes to population (around seventeen hundred breeding ewes according to the official statistics of 2023) and size (99–187 pounds [45–85 kilograms] for the ewes and 110–198 pounds [50–90

Figure 2.8. Värmland sheep.

Figure 2.9. Sylvester the Gestrike ram.

kilograms] for rams). The wool from Värmland sheep comes in different colors, from white through browns and to grays and spotted. I have seen the finest crimpy finull-type staples with mostly undercoat fibers, as well as long, cone-shaped rya-type staples, all in the same fleece.

Gestrike Sheep

Gestrike sheep can be white, brown, black, or gray and quite often spotted. Some are born black, but usually turn lighter with age. The most common staple type is the rya-type staple, sometimes with kemp, but I have also encountered a Gestrike fleece with mainly finull-type staples (figures I.1, 2.2, 2.19, and 4.2).

Helsinge Sheep

Helsinge (sometimes also Hälsinge) sheep can have a wide variety of both colors and patterns. Just like in many of the allmoge breeds, the wool from Helsinge sheep can be variegated over the body of one sheep, but the most common staple type is rya or gobeläng wool. The rams and some of the ewes have a collar of mane hair (figure II.1).

Svärdsjö Sheep

Svärdsjö sheep are generally white but can also be black or spotted. The wool is typically fine, crimpy, and soft, and it can have sort of a glittery shine that is rare in other breeds (figures 4.1, 4.3, and 4.6). Svärdsjö wool is known for its crimp that can sometimes be shaped like ringlets.

Dalapäls Sheep

Dalapäls wool has a particular shine. The wool is usually white, but it can also have gray spots. It is possible that there have been black Dalapäls sheep historically. The wool can be variegated, and it is not uncommon to find all four staple types in one individual fleece. The undercoat can be fine, but both mane hair and kemp can occur. Dalapäls sheep were used when the Rya sheep was reconstructed (above).

Klövsjö Sheep

Klövsjö sheep are usually white or black. Some can have spots on the face or on the feet. The wool is of rya type and can be long and lustrous (figures 2.16 and 6.3).

Figure 2.10. Dalapäls sheep in the shearing pen.

Tabacktorp Sheep

The smallest of the allmoge breeds, both in size and number (thirty-nine breeding ewes in eight flocks according to the 2023 statistics), is the Tabacktorp sheep. The ewes weigh around 44 pounds (20 kilograms) and the rams 66 pounds (30 kilograms). They are generally white, but around 10 percent of them are black. The wool can be variegated but is usually soft and wavy or curly (figure FM.1).

Fjällnäs Sheep

The northernmost allmoge sheep in Sweden is the Fjällnäs sheep. Their wool is usually white with a yellow tone. Some lambs are born fawn. The wool is usually of rya type in different lengths or variegated (figure III.2). Historically it has been common to divide the fleece into categories to suit different needs. Archaeological finds show that wool that is identical with the Fjällnäs wool in both color and texture has been used by the indigenous Sámi people.

Åsen Sheep

Åsen sheep have traditionally grazed in the forest during the summer and are hardy. They can have a wide variety of colors. Some are born black, sometimes with white or brown spots, and lighten with age. The wool can be variegated, but some can have the quite rare vadmal-type wool and is often easy to work with.

Roslag Sheep

The Roslag sheep have grazed the archipelago of Roslagen for centuries. Due to replacement efforts in the early twentieth century, there was a vast decrease in this breed. In the 1990s, there was only one flock left. Today the breed can be found in different parts of the country. Roslag sheep are usually black, white, or spotted. The wool can be described as long and straight and mainly of rya type (figure 4.7). Contrary to most of the other Swedish breeds, Roslag wool doesn't tend to felt. Historically, Roslag wool has been used for household needs in weaving,

Figure 2.11. Åsen forest sheep.

spinning, knitting, and nalbinding—to make anything from clothing and accessories to sails, and rya rugs.

Brännö Sheep

The newest allmoge breed was classified as such in 2023. Brännö sheep are generally white, black, or gray, sometimes spotted. The colors tend to lighten with age. The wool type is typically quite soft, but it can vary greatly from rya to finull type, sometimes with kemp or mane hair (figure 2.1). They stem from the island of Brännö in the Bohus archipelago, and the wool has traditionally been used for sweaters, socks, mittens, and rya rugs. Sometimes human hair has been added for strength.

Other Breeds in Sweden

SWEDISH LEICESTER SHEEP

Leicester Longwool sheep have been imported to Sweden at different times during the centuries, starting in the eighteenth century. In the beginning they were used to modernize the Swedish landrace. Larger flocks of Leicester sheep were imported from the mid-twentieth century. The breed has gradually been adapted to the Nordic characteristics when it comes to wool quality and skins. From the 1980s, the Swedish Leicester sheep have been bred with the same goals as the Gotland sheep—a homogeneous and shiny wool with large locks for the sake of lustrous

Figure 2.12. Brännö ewe lambs with the characteristic lamb locks at the chest.

Figure 2.13. Jämtland sheep, see also figure 1.1, left.

skins. Therefore, the wool of Swedish Leicester sheep is homogeneous, shiny, 4–6 inches (10–15 centimeters) long, with a micron count of 35–40 and with päls-type staples (figures 8.12–8.18 and figure 8.21).

JÄMTLAND SHEEP

Jämtland sheep is not a Swedish landrace, contrary to what the geographical name of the breed signals. It is a new breed, bred with the aim of getting a sheep in Sweden with wool soft enough for next-to-skin garments and as competition for imported Merino wool (see chapter 1). The breeding started in 2004 when Swedish Svea sheep (a cross between Texel and Finull sheep) were crossed with Danish Merino. The breeding took place mainly in the Swedish region of Jämtland, where the Svea sheep is common. The fleece is part of the breeding goals and should have a micron count below 23. Jämtland wool is popular for industrial purposes for garments as well as among hand spinners, due to the fine fibers.

POLL MERINO SHEEP

Through history Merino sheep have been imported to Sweden, beginning in the eighteenth century, with varying degrees of success. In the mid-nineteenth century, the competition from Australia became too hard. In 2020 Poll Merino embryos were imported to Sweden, and the first lambs were born in 2021. The vision of the Swedish Poll Merino Association is to decrease the import of Merino wool and, in the long run, make it possible for the whole manufacturing chain to take place in Sweden. Poll Merino has fine and crimpy fibers (figures 1.3 and 2.15).

Wool Characteristics

Wool around the world has a wide variety of characteristics. Local wool has been used for centuries, millennia, for purposes that suit that particular wool. Any characteristic a wool has is valuable and has a purpose; any characteristic has pros and cons in a specific context. An ultra-fine wool wouldn't be a good fit for rugs, just as a coarse wool wouldn't be a success in a next-to-skin garment. But the coarse wool would make a sturdy floor covering, and the ultra-fine wool would make the softest wedding shawl. As hand spinners we can look for the purpose that fits the wool we have locally.

The animal protection law says that sheep in Sweden must be shorn at least once a year. Many sheep farmers shear their sheep twice a year, though, usually in late autumn before the ram serves the ewes, and in the spring, a few weeks before lambing. The spring and autumn shorn fleeces can be of different qualities.

Spring-shorn fleece can be greasier and rougher than a fleece shorn in the autumn. The ewe may have produced extra lanolin and fine undercoat for extra warmth during the cold winter. She may also have been in gestation and a lot of the nutrients in her diet may have gone to the fetus. The fleece can be dirty and contain more vegetable matter from (indoor) feeding tables than the autumn-shorn fleece.

The sheep shorn in the autumn has usually been outdoors all summer and had access to fresh grass. The lambs may have been born a couple of weeks after the spring shearing, and the ewe may have kept most of the nutrients to herself. The fibers have usually grown longer during the summer than during the winter. An autumn shorn fleece can therefore be longer, cleaner, and of generally higher quality than the spring-shorn fleece.

Your local wool may come from totally different sheep breeds than the ones I have described above. Still, the characteristics of the fleece are what matter to me over the name of the breed. It is in the characteristics and, more important, the combination of the characteristics, I find what yarn and textile suit each unique fleece best.

Fineness

The diameter of the wool fiber is usually measured in microns. However, a micron count doesn't always show a fair picture of the fleece as a whole, especially in a breed with a variegated fleece. Sending a variegated fleece to a micron analysis can therefore be pointless. Comparing different fleeces to each other or comparing undercoat and outercoat from the same fleece can at least give an indication of fineness or coarseness.

Shine, Bulk, and Drape

The luster or shine of a fleece is usually connected to how the scales are placed on the fibers (see

Figure 2.14. Staples of spring-shorn Åsen wool from sheep not in gestation. There is more fine undercoat than in the autumn shearing. A quality like this is rare in spring-shorn fleece.

Figure 2.15. Gestrike wool of finull type (left), next to Swedish Poll Merino wool (right).

chapter 1). Scales placed edge to edge tend to be larger and reflect the light and present a shine, while scales wrapped over each other generally don't, and appear more matte. The shiny wool can be drapey through their smooth fiber surface (figures 8.12 and 8.13), while the matte wool can be bulky from the more textured surface.

Softness

Softness is connected to the fineness of the fibers, but that is not necessarily the only factor. The end-to-end positioning of the scales in the Swedish and other Scandinavian breeds gives the fibers low friction, resulting in flexible and adaptable textiles. The slipperiness of the fibers makes them feel soft. Even if they have a larger diameter than Merino fibers, fine fibers from Scandinavian breeds with their smooth surface can still feel as soft in a textile.

Length

A fiber's length can give you an indication of ways you can process it and ways it isn't advisable to process it. To get a yarn strong enough to work with, fiber must be twisted. A fiber needs the same number of twists to hold as a yarn regardless of the fiber length. Let's say a fiber needs twenty twists to hold. This means that both the 2-inch (5-centimeter) and the 10-inch (25-centimeter) fiber need twenty twists to hold as a yarn. While it can be possible to get twenty twists in the shorter fiber, the resulting yarn may be too tight to work as a yarn.

Fibers between 2 and 4 inches (5 and 10 centimeters) are fibers that I love to card. Short and fine fibers can make up the softest carded preparation and an airy yarn (see chapter 10). Since the fibers in a carded rolag have many connections in the jumbled arrangement, they will catch on to each other and hold as a yarn. Spinning fibers shorter than 4 inches (10 centimeters) from a combed preparation can be a challenge, since there is a risk that the aligned fibers glide apart in such a short length (see chapter 8).

Fibers longer than 4 inches (10 centimeters) are lovely to comb. The length of the fibers is well suited for the aligned arrangement of the fibers in a combed top. Longer fibers alone in carded rolags can wrap around the rolag and result in a tangled mess as I spin.

A combination of lengths can surpass the general recommendations of fiber length connected to preparation methods. A combination of shorter and longer fibers in a combed preparation won't be as strong as one with only longer fibers, but it will make the yarn less coarse and more airy. The same combination can also make wonderful rolags, where the shorter

Figure 2.16. Different fiber lengths in one staple of Klövsjö wool. The longest to the left are coarser and straighter while the shorter ones to the right are finer and crimpier. The image gives us a clue to the ratio of outercoat to undercoat.

Figure 2.17. From staple to finished yarn from a Gute lamb. Note the decrease of kemp especially after the teasing, but in the following stages as well. The hank has recycled sari silk that I have added at the carding stage.

fibers make up the airy foundation while the longer ones armor the structure, giving more strength to the resulting yarn.

Kemp

Kemp fibers (see chapter 1 and figures 1.2 and 2.17) are coarse and short, and don't conform in a yarn. They are common in some primitive breeds where they haven't been bred away. Some would say they are not desirable. However, they can play a significant part in a yarn. Their purpose for the sheep is to keep the staple open so that air comes in between the fibers and keeps the sheep warm. Let's translate that into a yarn: When you prepare a fleece with kemp in it, the kemp fibers will fall out sooner or later. You will quickly see kemp on the floor and in every corner of the house. This is good. As you spin and ply, the kemp fibers will push themselves out of the twist, just as they will when you knit or weave with the yarn. This is also good. So why work with a fleece with kemp if it's so good that it falls out? Well, it leaves room for something when it does. The kemp fibers take up a certain amount of space when they are in the yarn, but they leave air pockets as they push themselves out. And air means warmth. So, a textile with escaped kemp will be a light, airy, and warm textile. You see it already at the fleece stage; since the kemp fibers help keep the staples open and upright, there is no need for the undercoat to have crimp to stay erect. The armoring kemp has already taken care of that, leaving a light fleece. So, embrace your kemp. It will respond by escaping, leaving a lighter and warmer fabric in its wake.

Sometimes a fleece can contain kemp where it shouldn't. This can be because the farmer is inexperienced or hasn't considered the fleece in their breeding choices.

Evenness

Sometimes a fleece can have the same characteristics over the whole body of the sheep. This is common in modern breeds like the Merino, but also in some of the Swedish breeds like Gotland, Rya, and Finull sheep (see above). With a fleece like this a spinner can spin a uniform yarn from the whole fleece.

In other cases, the characteristics can vary over the breed, over a flock, or over the body of a single sheep, like in the case of many of the Swedish allmoge breeds (above). This means that the spinner can choose to spin a variety of yarns from the fleece of one individual if they sort the wool into categories (figure 4.6).

Willingness to Felt

Some wools tend to felt and others don't, and research isn't clear on what makes wool felt or not. Willingness to felt can be connected to how the scales are placed on the fiber. The edge-to-edge placement of the scales in a shiny wool suggests that fibers can slide in between each other and felt easily. The scales that are wrapped over each other give more friction and don't seem to felt as easily.

A yarn that felts when it shouldn't can be a disaster. But a fleece's willingness to felt can also give interesting opportunities. A fulled material is stronger and more windproof than an unfulled one (figure 5.2). A lightly fulled singles yarn can result in a straight stockinette fabric, while the unfulled and energized one will bias.

Some textile traditions include fulling, like nalbinding and two-end (or twined) knitting (figure 17.1). Sweden used to have water-driven fulling mills where people could come and full their woven textiles heavily into vadmal fabrics that would keep them warm through the winter (see the background textile of figure 5.2). There are still five mills left, some of which are used for educational purposes.

Crimp, Elasticity, and Stretch

Crimp is formed in the follicle (see chapter 1) and is unique for a breed and to some extent an individual. Since the outercoat fibers grow faster than the undercoat fibers, the crimp can help determine the difference in fiber fineness between staples when the eye can't; in general, fine fibers have more crimps per inch than coarse fibers.

The more crimp a fiber has, the more it can be stretched before it breaks. The elasticity shows how much the wool goes back after having been stretched. Wool is said to have a memory—it wants to go back into the shape it had on the sheep. The stretch and elasticity that come with high crimp will be a factor in the yarn you spin. If you want to spin a yarn with elasticity, perhaps for knitting a sweater, crimp can be desirable. If the fiber you spin into that knitting yarn has low or no crimp, you can still create elasticity to some degree by choosing an elastic knitting pattern like ribbing or cables. If you want to spin your yarn for a drapey fabric, you might want to choose fiber with less crimp.

Staple Formation

A staple consists of fibers from a small area of the body of the sheep. The shape of the staple can give us an idea about the content and ratio of undercoat and outercoat fibers—a staple with a wide base and a thin and narrow tip has a high ratio of undercoat and a lower ratio of outercoat (figure 2.14). A long and narrow staple can have mostly outercoat fibers, and a short and crimpy one with a wide tip can have mainly undercoat fibers. A fleece with staples that are equal in width from root to tip is usually a modern breed, like the Merino, where most fibers are similar to each other (figures 1.3 [left] and 2.15 [right]).

Strength

A coarser fiber is generally stronger than a finer one. The stronger fiber might be of interest if you want to create a strong yarn for socks or rugs or to blend into finer fibers to add some strength.

Changes in the environment or a sheep's nutrition can make the fibers thinner and weaker. If you see a change in color or quality over the length of a staple, this can indicate a weak spot.

Color

The wild sheep had a brown color. White sheep were developed quite early in sheep domestication, probably for the benefit of bright colors in dyeing. Many breeds today come in a wide variety of colors—from fawn to dark brown, from light gray to charcoal. Some breeds can lean toward red or honey; others are spotted. Shetland sheep come in many colors, color combinations, and markings, most of which have specific names.

If you look at a sheep, it can have a lighter color on the surface and a darker one underneath. This is usually due to sun bleaching. A fleece that has a big difference between the inner and outer part of the staples is something to take note of—if you keep a yarn from such a fleece in strong sunlight for a long time, chances are that it will bleach, too. In my experience, brown fleece bleaches more than gray or black fleeces.

Sometimes a difference in color along the length of the fibers has to do with the growing of the sheep. Gotland sheep, for instance, are usually born black and turn gray as they grow up. Some breeds can have a dark color or spots at birth that fade to almost white during their lifetime.

Finding a Fleece

In the section about wool characteristics above you have learned about the characteristics in their own beauty. It is in a context we can experience them as negative or positive. You will not hear me talk about a wool characteristic as negative on its own—there is a use for every kind of fleece. As hand spinners we have the opportunity, I may even say a responsibility, to find the characteristics and do them justice in a yarn. There is, however, such a thing as wool of questionable quality.

Where Should I Look?

First, to find a fleece—of any quality—you need to know where to look. As a spinner since 2011, I have built a network of sheep farmers I turn to when I am looking for a particular fleece. But there was a time when I didn't have a clue where to look. I got my first fleece on that first spinning lesson at the city farm. For a year or two that was my only fleece source.

One of the active members in the Swedish spinning group on Facebook was Kia, a Swedish wool classer who had worked with sheep and wool for many years in New Zealand and Norway. With her passion for rare breeds, she started a wool club in 2013 with fleece from Norwegian rare and endangered breeds, and I joined it. I call Kia and people like her a "node," a person who knows wool by hand and heart and knows where to find it.

Figure 2.18. Meeting the fleece (and the sheep owner) in person before you buy it is usually a good idea.

Figure 2.19. If I buy a fleece without having seen it, it is always from a sheep owner I know invests a lot of time in the wool as a product of its own. This Gestrike fleece was a gold medalist in the 2023 Swedish fleece championships that I bought at the online auction following the prize ceremony. I know the sheep owner, Claudia, and trust her completely.

In 2015, I went to the annual Swedish fleece championships. The fleeces in the championships were displayed in the competition hall for the visitors to look at. An event like this is an excellent opportunity to see different kinds of fleeces, different breeds, and variations within a breed. The medal-winning fleeces of this championship were auctioned off, and I won the auctions of the two fleeces that interested me. One of the sheep owners was at the venue, and I talked to her after I had collected my treasured auction wins. She told me about how she worked with her Rya and Finull sheep and their wool. Since then, she has won numerous medals, and I have bought many fleeces from her. I had text communication with the other sheep owner and learned about her approach to her sheep and their fleece. The wool from this breed, Dalapäls sheep, since then has become one of my favorites to spin.

The Swedish fleece championships has been the core of building my network of sheep farmers—keeping an eye on who wins the medals in the different breeds. Regardless of whether they win medals, these are sheep farmers who know about wool for crafting purposes. Some of them are even spinners themselves. Of course, the sheep always come first, but the fleece is present in the decisions they make every day, through breeding, selecting, feeding, and shearing. Claudia, mentioned at the beginning of this chapter, is also part of my network of sheep farmers, and I always turn to her when I want to explore the breeds she keeps.

In summary, look for fleece in the following places:

- Local sheep farms. Talk to the sheep owner and ask them about the wool.
- Fleece competitions or wool events. If you don't have one nearby, you might make a mini vacation of it. Perhaps go with another spinner or connect with spinners at the venue. Together you can

investigate the fleeces and get more information than had you done it alone. Talk to sheep farmers and ask about their flocks. The information you get at one event may be enough to provide you with a network of sheep owners to work with year after year.

- Spinning guilds or forums. There is so much knowledge and kindness in both Swedish and international spinning forums I have joined.
- "Nodes"—wool classers, brokers or agencies, shearers, spinning teachers—people whose wool judgment you trust fully.

Why Pay When I Can Get It for Free?

What about all that wool that sheep farmers sometimes dump (see chapter 1)? Or that your local sheep farmer offers for free? Of course there can be lovely fleece there. However, a high-quality fleece is rarely a lucky coincidence. Several factors contribute to the foundation of a high-quality fleece:

- Even nourishment for the sheep. Poor nourishment is, of course, bad for the sheep, but it can also cause weak spots in the fleece.
- There is as little vegetable matter in the fleece as possible.
- The sheep are dry when they are shorn.
- The space underneath the sheep at the shearing station is clean.
- The wool from the lower legs, belly, and tail area, sections shorter than 2 inches (5 centimeters), and felted parts have been removed.
- There is a minimum of second cuts (see below) in the fleece.
- The fleeces are stored separately in paper bags in a dry space.

A sheep farmer who sends their wool to spinning mills knows this: the higher the quality of the fleece the higher the yarn yield. And, of course, so do sheep farmers who submit their fleeces to fleece competitions.

What Am I Looking For?

For most of my spinning life, I have explored different Swedish sheep breeds. Now I can safely say I have spun yarn from all Swedish breeds. Quite often I have been looking for a specific breed, just to explore it. Sometimes, though, I am looking for particular characteristics for a project I have in mind. In 2020, I wanted to create a rya rug and kept my eyes open for a strong Rya fleece or a fleece with rya-type staples.

I can also look for a specific color. I have a soft spot for gray in all its shades. I also buy a lot of white fleece. I use white fleece when I teach and when I create videos since white wool is easier and clearer to see. For this reason, most of the wool in spinning photos in this book is white. A fleece of different colors, either spotted or variegated, is always interesting to me.

Sometimes I see a fleece and just know I need it. It can be a specific beauty, a combination of characteristics, or an extraordinary detail that I can't stay away from—perhaps a fleece that is very typical or equally atypical for its breed.

What I Avoid

The first thing I look for is stuff I don't want to see in a fleece:

- Vegetable matter. Vegetable matter (sometimes called vegetation matter) is stuff from the sheep's living room, both outdoors and indoors. It can be things in nature like seeds, leaves, needles, or soil. It can also be things from the barn, like hay, straw, or energy supplements. The vegetable matter can be superficial, between the fibers, or entangled. Superficial stuff is usually easy to remove. Stuff between the fibers can be removed through the different stages of preparation. Vegetable matter that is entangled between the fibers can be difficult to remove, though. This can include barbed seeds or burs. I will have to use force to detangle them, which can put a strain on both me and the fibers. I can take vegetable matter to some degree, but if the whole fleece is covered with stuff, I stay away from it.
- Poo. Usually, the person shearing the sheep removes the poopy bits from the fleece, but this isn't always a reality. If there is a lot of poo in a fleece, I would stay away from it.
- Second cuts. If the shearer goes over the same area of the sheep twice, the wool will be shorn twice. This leaves very short segments of wool cut on both ends, called second cuts. They can be between one-third and two-thirds of an inch (between 1 and 2 centimeters) and will turn up as nepps (small clusters of tangled wool) in the yarn and generally disturb the flow of the processing and spinning.

- Weak spots. Sometimes a fleece can have weak spots over the length of the staples. This can be due to environmental factors—there has been a shift in the environment or in the nutrition during a specific part of the sheep's life. Perhaps there is a difference in color or evenness of the staples. This could be a weak spot that breaks during the preparation of the wool. One important exception here is weakness because the sheep is molting. This is a natural characteristic of primitive breeds and not a sign of environmental factors (see chapter 1).
- Felted parts. The Swedish breeds felt easily, but if I see felted sections on the fleece, I may stay away from it. Sometimes the cut ends can be felted through the shearing. This can mean that there will be strain on me and on the fibers as I process the wool.
- Old wool. I want the fleece I buy to be freshly shorn, preferably from the latest autumn shearing (see above). Old wool can get brittle, which means broken fibers. The lanolin in old wool can get tougher to work through, and the fibers can be compacted through long storage. Wool that has been well stored and protected from moisture, insects, and rodents can often be spun after a few years.

Sometimes I can negotiate with myself. Perhaps there is a lot of vegetable matter or some felted parts, but there are at the same time qualities of the wool that appeal to me. In those cases, I need to draw a line between what I am willing to work through and what isn't worth the effort.

Local Connection

Connection with sheep owners is key; understanding their work gives me a deeper connection to the fleeces I use. I have had many conversations with sheep owners. I show them my appreciation of their fleeces and the hard work they invest in their flock, the pastures, and their 24/7 shepherding life. They tell me about the flock and their work with it. This is an ongoing conversation that we both enjoy. There is a lot in a fleece that I want to know but can't see, for example, when the fleece was shorn and for how many months the staples have grown, if shearing was done by a professional, or how old the sheep is. Anything that gives me a context about the sheep who gave me the fleece is important.

I also love to hear little stories about the sheep: the red stains on the side staples coming from the painted barn door Lotta the Svärdsjö sheep likes to scratch against (figure 4.3); how Pia-Lotta the Finull lamb was saved from slaughter at the last minute because she looked so much like her father who had been bullied to death so the sheep owner couldn't bear losing her, too; or Nehne, the Dalapäls lamb who was named after a girl who often came to visit the sheep farm. These stories mean something, and they help me connect to the sheep and their lives.

When I have bought a fleece, I always make sure to give feedback to the sheep farmer. They have provided me with stories and information about the fleece on the sheep, and I return the gift with stories of the fleece off the sheep. I tell them what I love about it. I show them what I have made from it and let them know that the yarn I have spun wouldn't have been possible without their hard work and dedication as sheep farmers. If there is something I don't love about the fleece I let them know that, too, and what it does to the wool when I work with it. The feedback I give will help the sheep farmer understand what I am looking for as a hand spinner.

Checklist for Buying Fleece

When I look at a fleece, I look for the things I want and the things I don't want (see above). I also do a few tests to make sure it is a good buy:

- I squeeze the wool to get a feeling of the bounce. If it pops back out as I let go, there is a good deal of bounce in it. This means that the finished yarn will take up space in the textile. A wool that doesn't bounce back may give more drape in the textile. Whether you want the drape or the bounce is up to you.
- I test the staples for their strength. I take a staple between my hands and tug it next to my ear. If I hear a pinging sound, it means that the fibers are strong. If instead there is a cracking sound, it means fibers are breaking. This can be an indication of a fleece in poor condition. If the fibers break in one particular spot in the fleece it may indicate a period in the sheep's life with a change in nutrition or of stress.
- I also test the fleece's willingness to be spun. I take a staple between my hands and draft out a small section of fibers from the cut end. As I draft, I twist the section into yarn, just as I would if one

of my hands were a spinning tool. I focus on how the fibers glide past each other. This is the draft I will face throughout the fleece. If the fibers are fighting me in this test, they will do so through all the steps of preparation and spinning (figure 2.20).

- I investigate the fiber types. A good way to detect the fiber types is to look at the shape of the staple. If it is conical or triangular, there is probably a mix of outercoat and undercoat fibers. If so, I can separate them to see their length and fineness. I hold the staple firmly with one hand on the cut ends (where there will be both undercoat and outercoat) and with the other hand on the tip end (where there will be only outercoat fibers). As I gently pull the fiber types apart, I wiggle the staple a bit. After the separation I can see how the staple is built (figures 12.4 and 12.5).
- And, of course, I look for the breeds, the characteristics, or combinations of characteristics that interest me.

Figure 2.20. To check the spinnability of a fleece I like to draft the fibers from the cut ends of the staple. That way my hands will get an idea what the wool may feel like to spin.

If you can't see the fleece before you buy it, make sure you see photos. Ask the seller to take pictures of different parts of the fleece, and in daylight. Ask the questions you need answered. There will be failed fleece purchases when you buy a fleece without having seen it in person; I have had them. But just as I did, you will learn from them. And by learning from your mistakes and by giving feedback to the seller you may get lucky the next time.

EXPLORE: CHAT

Wool can be strong and coarse or soft and warm. All types of wool have a purpose for the sheep, and there are uses for all types of wool. The wool that is local to you is worth exploring. Spinning local wool can enhance the feeling of connection to the wool and thereby the whole spinning experience.

If possible, find a fleece from a sheep that is local to you. Have a closer look at the differences you see. Is there a difference between longer and shorter fibers, other than the length? If you know the breed, read up on it. Does your fleece seem like the breed description, or is it different from it? Explore the fleece and see how it responds to your handling of it. Do you get inspired to work with the wool in a specific way–does a certain way of processing or spinning come to mind? Do you envision a specific textile technique or garment? You don't have to, but if you do, take note of it. What attracts you to this fleece? Have a chat with it!

What do you know about the fleece from before it was shorn? Where has the sheep grazed? Have you talked to the sheep owner? What do you know about the breed (or breeds if it comes from a cross)? When was the sheep shorn–in time and in season? How many months of growth? Ram, ewe, or lamb? Take notes and keep your notes close as you continue through the processing of the fleece in the upcoming chapters. You will thank yourself for this gift.

CHAPTER 3

On Washing Fleece

Through history, there have been various methods of washing, done in different phases—the washing of sheep, wool, or yarn. Pottery from ancient Greece tells the story of wool being washed in warm water before further processing. The Roman agricultural writer Columella describes the washing of fine-wooled sheep before shearing. He further advocates jacketed sheep (sheep with a cover to protect the fleece) to be uncovered every now and then, soaked in oil and wine, and washed three times every year in sunny weather. He also describes how wool is soaked after the dirt has been beaten out with sticks. Root of soap wort was used as a detergent. The village name of Shipbrook in eighth-century England refers to a stream in which sheep were washed. Traditional ganseys from Guernsey in the British Channel Islands were knit in unwashed yarn, since the lanolin helped keep the sweaters waterproof for the fishermen. Andean traditions include using a foam that is the result of washing quinoa grains from the husks, the high-altitude plant chukka, and fermented urine. Today most people in the area use laundry detergent or shampoo.

Figure 3.1. I buy raw fleece and wash it myself in water only.

In nineteenth-century Sweden, the washing of sheep and fleece occurred, but it seems like the norm in a large part of Sweden was to spin the wool in the grease and wash the yarn. Perhaps the spinner carded the unwashed wool by the fireplace, letting the heat melt the lanolin to allow for light and smooth movements. When I read about local traditions of washing either sheep or wool in Sweden, I see few signs of washing in anything but water.

In this chapter, I want to give a brief overview of fleece washing so you can find a way that suits you in your climate, your type of water, and your type of wool, in the space and with the equipment you have.

Why Wash??

In chapter 1, we talked about the wool grease, lanolin, and the sweat, suint. Both substances come from glands that are part of the follicle where the fibers originate and grow and, thus, coat each fiber. The combination of lanolin and suint—yolk—keeps the fibers elastic and smooth.

Dirt and vegetable matter are other substances in the wool, but they come from the sheep's environment. Let's have a closer look at which of these fleece ingredients we can wash off, which we want to wash off, and how we can do it.

- Suint has no purpose in the spinning process. It is water soluble and will be removed in warm water. Suint can help get the fleece clean (see below).
- Lanolin, the wool grease, is not water soluble. It will, however, melt at 104–111°F (40–44°C). Some spinners want some lanolin left for spinning. Soaking in warm water may therefore be enough. Others want to remove all the lanolin. For this, add a washing agent.

- Dirt has no function in the spinning process and is something we usually want to remove. It can often be bound to the grease. To remove the bound dirt, we therefore need to detach it from the grease. When the lanolin melts, the dirt will be released and cleaned off the fleece.
- Vegetable matter—seeds, needles, grass, peat, hay, and straw—will create tangles in the wool, disrupt the flow of both wool preparation and spinning, and may lead to an uneven yarn and more waste. It is not soluble in either water or washing agents. Therefore, it needs to be removed in ways other than washing (see below and chapter 4).

Out of these four ingredients—two from the sheep and two from the environment—we want to remove three. The fourth, the lanolin, is a matter of personal preference.

Washing, *soaking*, and *scouring* are words that are used for cleaning fleece. We need to define these to know what we are talking about. Let's see *washing* as the mother word, for the concept of, to one degree or another, cleaning a fleece (or yarn). Scouring is a way to wash thoroughly, with the aim of removing all dirt, suint, and lanolin. Hand spinners who want to remove all the lanolin will need to scour the fleece. All commercially prepared wool is scoured before it goes through the machinery, since the lanolin can damage and ruin the equipment. If the wool is going to be dyed, it also needs scouring for the dye to set properly. Soaking, at the other end of the washing spectrum, is to place the fleece under water and leave it to do its thing. I want some lanolin left when I spin. For this reason, I usually soak my fleece—to get rid of suint and dirt but keep some of the lanolin. Washing after spinning—as has been common in parts of Sweden historically (see above)—is also an option. Spinning unwashed wool is usually called spinning in the grease, although some spinners refer to spinning with any amount of lanolin left in the wool (unwashed or soaked) as spinning in the grease. For the purpose of this book, I define spinning in the grease as spinning unwashed wool.

For many reasons, a hand spinner may choose to remove all lanolin. It may be that they don't like the smell or feel of it, or that it doesn't feel clean. Some people can get allergic reactions to lanolin. Lanolin can react with surrounding substances and shorten the shelf life of the fleece. A fleece stripped of all its lanolin can therefore usually be stored for a longer time without deteriorating.

The spinner may choose to keep lanolin in the fleece for many reasons. The spinner might enjoy the slow and smooth pace it creates in both wool preparation and spinning. A fleece stripped of lanolin can feel brittle and, to some, lifeless. Working with lanolin in the fleece can do wonders for the hands—the lanolin, after all, is made to protect the skin (see chapter 1). The lanolin can also be a reminder of the living animal that grew the fleece. A yarn that is spun with lanolin in it gets lighter after washing. The removal of the grease from the yarn gives room for air, which makes the yarn warmer.

Factors to Consider

I can't tell you how to wash your fleece; there are too many factors to consider. I can only tell you about different methods and how I like to wash my locally acquired fleece according to my preferences and with the conditions I have in my home.

How and when you wash your fleece is up to you. There are several factors you need to consider:

- First, you need to decide how you want to work with your project, and from there decide when to wash. Do you want to wash your raw fleece or spin it unwashed and wash the yarn?
- You also need to decide if you want to remove all the lanolin or only some of it. If you want to dye the wool before spinning and get a clear color, you need to remove all the lanolin. If you prefer a more variegated dye result, you may choose to leave some lanolin.
- Related to the amount of lanolin to remove is the amount and quality of the lanolin in the fleece you work with. In general, wool from fine-fibered breeds has a high amount of lanolin (that can also feel waxier than in other breeds), while wools from longwool breeds, landraces, and primitive breeds are low in lanolin. Medium-fiber breeds and crosses usually fall somewhere in between.
- The season the sheep was shorn is a factor, too. A fleece shorn in the spring generally has more lanolin than a fleece shorn in autumn, at least in the Northern Hemisphere (figure 5.1).
- The hardness of your water is also a factor. Hard water may need more washing agent than soft water, or a water softener.

- If your pipes are sensitive to the grease, you may want to consider discarding the water outdoors. If you use a washing agent, it is extra important to choose one that is organic and biodegradable. You may also consider washing one fleece at a time.
- What equipment do you have? If you don't have a vessel that can accommodate a whole fleece, you may want to wash in sections.
- You can wash your fleece as it is or in a mesh bag or large pillowcase that is big enough not to compress the fleece. Lifting the wool in and out of the tub is easier in a mesh bag, but in my experience, washing without a bag makes it easier for any second cuts (see chapter 2) to run out with the wastewater.
- Do you have a washing machine with a spin-only cycle? If you want to spin cycle your fleece as a first drying step (see below), test with a sample of your fleece first to make sure it doesn't felt when you do that.
- Do you have solutions for drying your fleece (see below)? It may take a few days, depending on the humidity and how airily the wool is placed.
- Washing and drying wool tend to take up space. Consider the space you have, both indoors and outdoors, and what people or pets share your space.

General advice about washing fleece, regardless of method:

- First, try different methods and different ways of working, to find what suits you best (see below). Wash small batches and compare the results and the workloads.
- Add the fleece to the water (as opposed to adding water to the fleece) to avoid felting.
- Leave the fleece in the water when you have pushed it down. Any agitation will risk tangling or felting the wool.
- If you use a washing agent, make sure it has a neutral pH. Wool is sensitive to alkalinity, as well as to excessive acidity (not generally an issue in home processing). You may want to add some white vinegar in the last rinsing water to neutralize any alkalinity.
- When you change waters, make sure the new water matches the temperature of the water from which you removed the fleece. A temperature shift can felt the wool, especially if a washing agent is involved.

If you don't use any washing agent (or use one that is safe for plants), you can use the wastewater as fertilization for your plants. A more concentrated waste, like the fermented suint water (see below), needs to be diluted, though, with up to nine parts water.

Ways to Wash Fleece

There are many ways to wash wool, and whoever you ask about their wool-washing method will give you a new answer. Washing can also be done at different stages of the process.

As a first step before washing, some people prefer to remove unwanted parts and sort the fleece. I am usually eager to wash the fleece as soon as it comes into the house, though. Therefore, I do these parts after the washing, when I pick my fleece (see chapter 4). If you choose to remove unwanted bits before you wash the fleece, spread the fleece out on the floor or ground (preferably on a tarp or at least not straight onto the ground). If possible, lay out the fleece as it grew on the sheep, cut ends down and as much as possible in its entirety. This way you can see what kind of staples grew where on the sheep. Remove large pieces of vegetable matter, poo, second cuts, and felted sections. This is also a time to decide if you want to make a rough sorting of your fleece into sections like color, staple type, or staple length. You can read more about sorting the fleece in chapter 4.

Use a mesh bag for your fleece or don't. Bagging your fleece may be a good idea, especially if you have sorted it into sections. Some people like to gently bundle their staples before washing to keep them organized. This may be a good idea if you are working with long staples of very fine wool, like a whole year's growth of a Merino or Merino cross fleece. Arrange the staples with all cut ends in one direction and all tip ends in the other. Either tie them loosely into small bundles with string or roll them loosely into a rectangular piece of cotton fabric to keep them in order.

Most articles I've read about using a washing agent suggest a dishwashing detergent rather than a laundry detergent. Soap is more environmentally friendly than dishwashing detergent and will work, too, if you wash in soft water. There are also washing agents on the market specifically formulated for cleaning fleeces.

To wash your fleece, with or without a washing agent, fill your tub with water that is at least 111°F (44°C). To scour, the water should be at least 140°F

(60°C). Either way, use rubber gloves to protect your hands. Add any washing agent first, give the water a stir, and then gently push your fleece down. Some people who use a washing agent make a pre-soak first, without the washing agent, and follow up with one or two soaks with the washing agent. The pre-soak will allow the suint and lanolin to loosen and the fibers to take up the water in the upcoming washing agent soak(s) more easily.

Lift out the wool and rinse it in additional soaks until the water is nearly clean. In any of the baths, let the wool soak for a maximum of twenty minutes. The water temperature shouldn't sink below 104°F (40°C) or the lanolin will cluster back onto the wool.

If you don't use the spin cycle of your washing machine as a first drying method (see below), gently squeeze out the water with your hands or roll the wool into a towel and stand on it.

The Fermented Suint Method

In chapter 1, we explained that the lanolin and the suint in the fleece create a natural soap that cleans the fleece with the help of rain and the sheep's heat generation. This process is also an important factor in the washing itself. When we clean the fleece, we can take advantage of the sheep's natural ability to clean its fleece on the hoof.

Figure 3.2. The bubbles in the water come from the lanolin and suint in the wool. There is no washing agent added.

The fermented suint method doesn't require an external washing agent and uses less water. It is therefore a method that is usually more environmentally friendly than other methods, assuming that the fleece hasn't taken up any chemical substances. It doesn't remove all the lanolin, but if the spinner is aiming for a lanolin-free fleece, it requires less washing agent compared to other methods.

The fermented suint method works for cleaning a larger number of fleeces in a row. I have washed between five and ten fleeces this way, in the same water, from May to November, and it's the method I use when I have multiple fleeces to wash. The method requires outdoor space.

To start, choose the dirtiest fleece you have, and soak it in a tub of water outdoors for five to seven days. Room temperature is ideal, so the summer is the best time to do it. This dirty fleece is your starting fleece, like a sourdough starter.

During this soaking period, the water will start to ferment. It will not smell like raspberry pie, so do keep a lid on the tub, for your sake as well as your neighbors'. There may be a film on the surface of the water, probably also a few soapy bubbles. This is the suint doing its thing. Lift the fleece from the fermented water but keep the water! Rinse the fleece in clean water until it is clear, and then dry it (see below).

Soak your next fleece in the fermented water for a couple of days and rinse. Keep soaking your fleeces one at a time in that same starting water. It will get gunkier, smellier, and more potent as you go, just as it should.

As you will see, the fleeces will get magically clean in the dark brown and appalling liquid. The fermenting soak won't remove all the lanolin, though. If you are aiming for a lanolin-free fleece, you can add a washing agent to your first post-soaking water, but you will need far less than for a scour.

The fermented suint method isn't for everyone. The smell can be intrusive to say the least, and if you are worried about potential bacteria growth or contamination with traces of parasite treatments, you may want to use a different method.

How I Wash

The Swedish landrace and allmoge breeds (see chapter 2) that I work with are quite low in lanolin. I prefer having some lanolin left as I spin. For this reason, I never use any detergent when I wash wool from these breeds (which is most of the time). When I have worked with wool from the fine-fibered Swedish Jämtland sheep, which has some Merino in it (see chapter 2), I have scoured the wool to remove all of the lanolin.

When I buy fleece, I make sure it is raw. I want to control the washing process myself to get the result just the way I want it.

This is how I wash most of my fleeces:

- As soon as I get a fleece home, I wash it. I fill a large tub with the warmest water my tap provides (around 131°F [55°C]) and put the fleece in the water. If I see large pieces of vegetable matter, I remove them.
- I leave the fleece in the water for around fifteen minutes; it mustn't cool down too much.
- I remove the wool and pour the water out. Typically, it has the color of weak coffee.
- I refill the tub with water of the same temperature as the previous soak and repeat the process for a total of perhaps three times or until the water is nearly clean.

After this process, the suint and the dirt have most probably run out with the wastewater. Some lanolin has melted and run out, too, but some is still left. Most of the vegetable matter is still in the wool and will be removed through the steps of wool preparation (see chapters 4, 7, and 8).

If I wash wool in the summer, I like to do it outdoors, using rainwater from our rain barrel. I take the same steps as I do indoors, only the water is cold, and I can leave the fleece in the water for a day or two. This may not melt the lanolin, but the soaping process from the lanolin and the suint may still clean the dirt off. I may even see some soapy bubbles in the bath. All that is left is the clean fleece with some lanolin and any vegetable matter that was there to begin with.

When I wash my finished yarn (whether the fleece has been washed or not), I soak it in warm water with an organic shampoo. In the last water I add some white vinegar to neutralize any alkalinity.

How I Dry Fleece

After the soaks, the fleece needs to dry. As a first stage, I spin it in the spin cycle of my washing machine. This may sound like felting doom, but it is not. Usually. While the drum spins, the wool lies still against the walls of it, and water and dirt are spun out. Beware, though, and test drive the wool before you do this. Twice through my wool washing career my fleeces have come out not felted, but compacted. Luckily, I have been able to save them.

Finding space for drying may involve some innovation and negotiation with the space in your home and any people or pets living there.

- In the summer, I like to dry my fleece outdoors to let the wind blow through it. Sometimes I place a compost grid over the seats of two garden chairs, place the fleece on the grid and another grid on top, like a fleece sandwich.
- I also have a few mushroom trays from the supermarket that I store on top of each other for indoor or outdoor drying. The upright storing is perfect for a balcony.
- Sometimes I dry my wool in mesh bags on the clothesline and let the air flow through.
- For indoor drying, placing the wool on newspapers underneath the heat pump or in front of the fireplace (not too close, though) is a good idea.
- Sometimes I put a compost grid on top of four egg cartons, one in each corner, and place the fleece on the grid. If possible, I slide the contraption underneath a table and hope my family won't notice.
- I like to fluff the drying fleece now and again, being careful not to disrupt the staple structure. This helps the drying process, and some vegetable matter will fall out.

When I have washed wool with water only, I try to prepare and spin it within a year from the shearing date. Old fleece can get brittle and the lanolin stiff. Every now and then I go through my fleece stash and check the freshness.

Troubleshooting

As in any step of the process from fleece to yarn, things can go wrong. I see these occasions as opportunities to learn and get to know the fleece I spin. Here are some suggestions to check should your washing have unwanted results.

- **My fine-fibered fleece is still tacky and sticky.** Perhaps this fleece needs two soaks with a washing agent. Grease can also vary, and different washing agents can work for different kinds of grease. Try using two different washing agents in two different baths. Another reason can be that the water has cooled off and the lanolin redeposited on the fiber. This is more common with fine fibers and can be difficult to fix.

- **My fleece is tangled or slightly felted.** A felted fleece will not unfelt, but sometimes it is salvageable. Picking the fleece (see chapter 4) is essential for opening up a tangled or lightly felted fleece. It may require some muscle power, though. For your next wash, try using mesh bags. It makes it easier to lift the bag out of the soaks without handling the wool. Also make sure the water temperature is even across the soaks.
- **The lanolin is clumped in the wool.** Have you let the water cool down with the wool in it? If so, the melted lanolin may have gone back to its solid state and reattached to the wool. Make sure the water temperature doesn't sink below 37°F (43°C).
- **My wool feels harsh.** Check your water hardness. If your water is hard there are softeners on the market. You can also try a different washing agent or use a larger amount.
- **My wool feels brittle.** This can happen when all the lanolin is removed. Try washing off less lanolin next time and see if it feels better.
- **My washed and dried wool has lots of static electricity and is difficult to prepare.** This can happen with wool that has been stripped of all its lanolin, especially in a dry climate. For easier wool preparation, try using a spinning oil. In the winter when my indoor air is very dry, I like to spray my wool with a combination of 2 tablespoons of water, 1 teaspoon of coconut oil, and a drop of dish washing detergent. Since the coconut oil is solid at room temperature, I warm the bottle in my armpit to melt it before I use it.

EXPLORE: WASH

Wash your fleece, or part of it, in a manner that works for you and the space and material you have available. To compare, try a few different methods to see what suits your space and your preferences. Save a few staples and note the difference between the methods, and between washed and unwashed wool. You can also spin from your different preparations (and from the unwashed sample) and note the difference. It may take a few times to find what suits you, and you may choose different methods for different breeds or occasions, but give it time. Explore to find a washing method that works for you.

CHAPTER 4

Staple by Staple: Pick Your Fleece

Picking a fleece means pulling out one staple at a time from it. It is well-invested time to explore the fleece and experience its characteristics. In this chapter, I want you to learn about the benefits of picking a fleece, from ergonomics, waste, and pre-preparation to the opportunity to sort the fleece according to the characteristics you experience as you work your way through it, staple by staple.

Figure 4.1. Picking the fleece is the first step of the journey from destroying the structure the sheep has built as a fleece to creating your own as a yarn and textile.

Bring in the Air

When I get a fleece, it's not much different from when it was shorn off the sheep. This is a perfect opportunity for me to spend some time with the wool and get a first glimpse of what it is about. Picking is also a way to prepare for and ease upcoming steps. Yet it is so much more, and I will get to that, too. If I am planning to spin the wool washed, I generally wash before I pick (see chapter 3). That way the air I bring in between the fibers stays in.

Preferably I pick my fleece outdoors if the weather allows it. If not, I do it indoors with newspaper underneath the fleece. To begin picking the fleece, I grab a piece of it in my nondominant hand. With my dominant hand, I search for the tip ends. Usually, they are a bit denser than the cut ends, perhaps pointy. Sometimes, especially if the sheep has been jacketed, the

Figure 4.2. A washed and dried fleece, ready to be picked. The basket is a traditional Gotlandish saigkorg, used for storing carded wool.

Figure 4.3. Picked (left) and unpicked (right) wool from Lotta the Svärdsjö sheep. The air is the difference. The tint of red comes from the barn door she likes to scratch against.

Figure 4.4. Tip ends and cut ends straight off a newly shorn Dalapäls sheep. The cut ends are clean and airy; the tip ends, which have been facing the elements for a longer time, are dirtier and more compacted.

tips are a bit matted. I hold the cut end surface of the fleece firmly and pull out the staple by the tip end. I pick the next and the next staple. I do what I can to pick only one staple at a time, even if the staples are very thin. Sometimes it can be difficult to see where one staple begins and another ends, but I do my best. This is not an exact science. Rather, it is a way to dissect a fleece and the first step on the journey from fleece to yarn.

When I pick the staples, I invite air in between the fibers for the first time after the shearing—not much, but enough to make the upcoming steps easier. Just the fact that the cut ends of the staples are separated from their neighbors will make the steps of teasing and carding easier—on my body and on the wool. Sometimes the cut ends have felted slightly through the shearing, which makes picking the staples even more beneficial. Without picking the locks before teasing (see chapter 7), I would have to use more force. This is force that would strain my body and the fibers. Fibers that break in the preparation are fibers that go to waste (from a spinning perspective—wool on the compost and as fertilizer in the garden is not wasted to me). In this sense, picking will reduce the proportion of waste.

Bring out the Stuff

As I invite air into the wool, stuff will come out. Either by my hand or simply by being replaced by air. As I go through the fleece, I remove any larger vegetable matter I find. It can be from feeding, from shelter, or from nature—hay, straw, seeds, needles, and other pieces from the sheep's living room. Removing vegetable matter at this stage is something I will thank myself for later. These are pieces that otherwise will make carding or combing a lot more troublesome because they will obstruct the flow of the processing and, again, put unnecessary strain on my body and on the fibers.

If I look at the floor or surface underneath the fleece, I will see a carpet of vegetable matter that has loosened as I have been picking along. If the fleece has kemp, some of it will fall out during the picking, too.

Other things I can remove include felted parts, coarse wool, brittle fibers, second cuts, and poo (see chapter 2). Any matter that I don't want in my yarn is matter that I remove from the wool at the picking stage. All of this is visible to me since every single staple goes through my hands, right in front of my eyes.

What to keep and what to toss is a matter of taste, time, and prioritizing. Small seeds and vegetable matter that are softly entangled in the wool will fall out in the picking, teasing, and carding or combing stages. Sometimes vegetable matter is deeply entwined in the fibers, though. In situations like these, I need to decide whether to keep the fiber or put it on the waste pile.

Second cuts (see chapter 2) are usually located at the cut ends of the staples and are easy to see and remove as I pick. Second cuts left in the wool in further steps of the process will result in nepps and uneven spinning.

If I work with a primitive fleece that sheds (see chapter 1), picking is a perfect opportunity to find the rise, where the fibers are weakened or broken to give way for the new year's growth. I hold the fiber mass, perhaps a little firmer than with nonshedding fleeces, and pick the staple, hoping that it will break at the rise. This is usually only a centimeter or so from the cut ends, depending on how long after the rise the sheep has been shorn.

What I See

When I go through the fleece, I notice things. Everything I see is information that can help me make decisions along the way.

I may see different kinds of staples and staple constructions on different parts of the fleece (figure 2.2). Many of the Swedish heritage breeds can be variegated with sometimes four different staple types in one single fleece (see chapter 2). I take note of this, too. With a homogeneous fleece, I can make a larger project with the same kind of yarn. With a fleece with different kinds of staple types, I can spin different kinds of yarns. This is an opportunity that is unique to me as a hand spinner—no spinning mill will separate a fleece into sections depending on individual staple types. It warms my heart that I can do whatever I want with the wool, depending on its differences, and through the fact that my hands are in the fleece. In surveying the fleece, I can also detect different qualities in different parts of it. The staples may be built in a similar manner, but with coarser or finer fibers. I take note of this, too.

As I pick a staple, I see how it is constructed—outercoat only, undercoat only, or a combination of both. I also see the length of the fibers. With all this

Figure 4.5. Most of the Swedish breeds have dual coats with a visible difference between outercoat (center) and undercoat (right). In this example of Värmland wool, the fiber types also have different colors.

information, I can get a first idea of how I want to prepare and spin it. If I see only fine undercoat, I may want to card it and spin woolen into a soft and warm yarn. If I see only long outercoat, I may lean toward combing it and spinning it worsted into a strong and shiny yarn (see chapter 10). A combination of outercoat and undercoat can leave me several choices: card or comb it together, separate and prepare differently, or semi-separate to create a custom yarn with the undercoat to outercoat ratio that suits an idea of the yarn I want to explore.

I may see kemp in the wool (see chapter 1). This might be seen as a negative thing, but it doesn't have to be. Kemp facilitates an open wool, giving warmth to the yarn. Through all the steps of the process, the kemp will fall out, on its own as air comes in, and by the motion of the tools. Kemp fibers falling out of the wool leave air pockets. Thus, the presence of the kemp during preparation and spinning can make the finished yarn softer and more open as it falls out.

I may see color. As a hand spinner, I have the advantage of exploring how the color varies. Are there different colors in spots across the fleece (figure 2.12)? Do the outercoat and undercoat fibers have different colors (figure 4.5), or are there different colors across the length of one fiber? Perhaps it is even a combination of all of these.

What I Feel

With my hands in the fleece, I sense what it is about. As I pull a staple out of the mass of staples I experience how the fibers glide past each other, how they relate to one another, and how the lanolin feels against my hands. I get an understanding of the elasticity and the bounce of the wool and how different kinds of staples may feel different.

As I go through the wool, I get to know how the staples and the fibers work in their construction and in movement. Does the staple I pick separate easily from the fiber mass? Do I need to tug, or don't the staples hold together at all?

In this first experience of the fleece as it was shorn off the sheep, I get an understanding of what it may feel like to process and spin this wool. I like to see it as a warm-up for my hands and my brain in what is to come. What's more, my hands store this information (see chapter 18). As I work my way through the fleece, my hands learn, over and over, how the fibers respond to my handling of them. If I pull a few strands from the cut end of a staple and draft them out, lightly twisting, I get a sense of how the wool may feel to spin later (see chapter 2 and figure 2.20). Do the fibers glide smoothly past each other, or are they fighting me? Are they silky, or do they give some sort of resistance? How do they feel against my hands as I draft? All this information will help me make decisions along the way. It's like I ask the wool questions by handling it, and the wool replies in its reaction. The conversation that unfolds between hands and wool is to me at the core of processing and spinning.

If I come back to this fleece later, my hands will recognize it, like an old friend—perhaps awkwardly at first but, after a while, like we had never been apart. Again, my hands are key here through their ability to interpret the wool in a way that a machine never can. We will get back to this in chapter 18.

Establishing a Relationship

Digging my hands into the fleece, whose previous home was a living being, connects me to that being. The fleece of a sheep grazing on the green hills of Shetland may reveal small granules of peats or sections of miniature moss, while one of a forest grazing sheep may surprise me with juniper needles or barbed seed pods grateful for the free ride.

By investigating the staple along its length, I can see deviations along it. A change in quality can indicate a change in nutrition between seasons or be a sign of pregnancy and lactation. Perhaps I know something about the sheep owner and the pastures; perhaps I know the name of the sheep. All of this keeps me connected to the sheep and grounded in the notion of the fleece as protection and part of a living being.

If I take one step at a time in the preparation of the fleece, I will have a clear connection to the previous step and thus all the way back to the sheep. This connection is important to me; it reminds me that the wool is a gift and that I have a responsibility to give it my best to make it shine.

It is possible, of course, to pick the fleece with a wool picker, a mechanical tool that picks larger sections than you can by hand, which would be a lot faster. Just as picking by hand, a wool picker opens up the fleece and brings air in and vegetable matter out—though probably not with the same thorough result as with hand picking and probably with more waste. And without getting to know the fleece. Still, it's a possibility.

Thoughts on Sorting

As I pick my wool, I take the opportunity to survey it and decide if I want to sort it into categories. From these I can choose to prepare and spin a range of different yarns.

The most visually obvious category to sort into is color. Perhaps the fleece is spotted with clear sections in different colors. I can roughly divide the fleece into color piles and pick the staples from the individual piles. Sometimes the color variation can be within a staple, with different colors in the outercoat and undercoat. Perhaps I want to separate the fiber types for that reason. Or, as sometimes happens, the color varies over the length of the staple. I can't really sort the top from the bottom of a single staple, but it is still something to take note of as I plan how I want to prepare and spin it.

Fiber type is another category I use to sort the staples during the picking stage. If I want to spin outercoat and undercoat differently, I can separate the coats by hand (see chapter 3 and figures 12.4 and 12.5). This way I can investigate the characteristics of the different coats in the process. Sometimes I decide to sort according to both color and fiber type—outercoat in different color piles and undercoat in other color piles (figure 6.3). The possibilities are endless.

If a fleece has different staple types, I can sort the staples into the categories I see and prepare and spin different yarns from those different categories. Imagine a crofter in the 1850s with a flock of perhaps only five sheep, and the range of yarn for different purposes the sheep owner would be able to spin for the family with a breed with a variegated fleece.

Perhaps I sort my fleece according to a planned end product—long and strong fibers for weaving, short and soft fibers for next-to-skin garments, or a combination of both for something that needs both warmth and strength, like a pair of mittens.

Figure 4.6. This (rather uncharacteristic) Svärdsjö fleece has a few different staple types, and I decided to sort it into a few categories.

Figure 4.7. I store my wool in cardboard boxes or paper bags with breed (and sheep name if I have it), weight, and shearing date. These were all shorn in the autumn of 2023.

When I am done picking my fleece, whether I have sorted it or not, I have not only a clean and airy wool but also a gift to myself. I have laid the table for the upcoming preparation of the wool, making it easier on my body and on the wool when I comb or card it. Furthermore, I have gotten to know the fleece, literally staple by staple. At this stage, as another gift to myself, I can make notes of the characteristics of the wool and my thoughts on how I may prepare and spin it.

Storing

Usually I do not prepare and spin a fleece directly after having picked it. I have a queue of fleeces in my storage and try to keep a strict system—first in, first out. So, when I have picked my fleece, it is ready for storing, unless I cheat (which I sometimes do). I place the fleece in a paper bag or cardboard box and write the breed, the name of the sheep if I know it, and what year and season it was shorn on the container. If I have sorted the staples into categories I roll each category into newspaper or keep them in separate bags.

Ideally, I would keep my bags of picked fleece in a dark and cool place, but that is not possible in my reality. I keep my fleece bags and boxes in the compartment underneath our sofa bed. That way they are out of the way and hopefully out of reach for moths and other pests.

EXPLORE: PICK YOUR FLEECE

If you haven't already, take a piece of a fleece and start picking it, staple by staple. Give yourself the time and the space to be in the fleece. Sit comfortably, make sure the lighting is good and that you have a plan for where to keep picked and unpicked wool as you work your way through the fleece. I like to pick for an hour or so in the evening for a few days or even a week. If weather and temperature allow, I work outdoors.

Perhaps you have observed differences in the fleece that you want to sort: color, fiber type, staple type, fiber length, and/or wool quality. Read up on wool characteristics in chapter 2 if you like. If you decide to sort your fleece into categories, do that. Store the different categories in separate bags or rolls of newspaper and mark them.

As you pick through your fleece, observe what you see and what you feel. Make notes of the decisions you make and what thoughts about the fleece emerge. Notes can include how you experience the fibers as you move through the fleece. Perhaps you continue a thought you started while completing the "Explore: Chat" exercise in chapter 2, about how you would like to prepare it and what kind of yarn you think it would like to become. Nothing is wrong here. Anything and everything you think of is information you will thank yourself later for having written down.

Create and Destroy

Before me is a raw fleece—dirt, lanolin, vegetable matter, and all. The staples display a combination of characteristics that has protected the sheep, millimeter by millimeter. The tips, long and dense, hang out from the fleece like rats' tails. The undercoat, billowing, keeps the outercoat fibers upright; the outercoat, armoring at the base, allows air to come in. Even though the fiber types have individual tasks, they work together for the sheep.

As I get the fleece in my hands, I marvel at the wonder of it. So much information can be unfolded in this coat; to learn and to play with. In this original state, I give myself the time to study how the fleece has worked to serve the sheep. How are the staples put together? Where have different types of staples grown on the body of the sheep? How can I take advantage of the characteristics of the fleece and make for me what the fleece once was to the sheep?

I need to destroy it, is my answer. To create something new I need to destroy, disassemble, and deconstruct the fleece. In each step, I take measures to investigate and learn from the characteristics of the wool. Sheep have grown their coats for millennia. Who am I to question their ability to create a high-quality product?

Perhaps I see colors, length, fiber type, and every thinkable expression of curl. I may feel any degree of silkiness, friction, and bounce. All these characteristics are there for me as a compass on the road to my coat. I can choose a few to play with, others to highlight. I may divide into categories or let the differences work together. But I will destroy their original shape and composition.

By inviting air between the fibers, I separate them. After teasing and carding, the fleece is unrecognizable. The air the sheep breathes—the air that the fleece once protected the sheep against and that nestled between the undercoat fibers to keep the sheep warm—is the same air I use now to destroy that protective structure.

Step by step I build a new coat. Perhaps I work the fiber types differently—card the fine fibers for warmth, comb the strong ones to stand united against the rain. And, in the next step, perhaps I spin extra air into the carded wool and keep the combed dense. Finally, I may weave the very different yarns together into a twill, allowing them to cooperate again. On the weft side soft and warm against my skin, on the warp side strong and shiny, bouncing the rain off.

Through destroying the coat that the sheep needed to be freed of, I create a new one for me, a textile that honors the treasure the sheep grew. All the while the sheep is grazing and growing itself a brand-new coat.

Figure II.1. Examples of different stages of a Helsinge fleece from raw fleece to woven and knitted swatches.

CHAPTER 5

Superpowers

I am shearing Parisa, a two-year-old Dalapäls ewe in Lena's flock. The newly sharpened shears glisten in the April sun, as I look for a pocket in the fleece to insert the tips. I take a deep breath and do it. Fiddling, I shear a path along her spine and make a space to move from. Lena, the sheep owner and an experienced spinner, shears Ester on the shearing table opposite mine. She has shorn these sheep twice a year for eighteen years now. I am at most a beginner. As I move across Parisa's body, I see what her fleece is like; her body presents a map of fleece qualities, from the weathered strip along her spine, the cascading staples on her sides, and the miniature around her neck. For every staple I shear off her body, I learn how the fibers work together to shield her from the elements.

This is as close as I get; this is where no mechanics and no external process stand between me and the wool. This is the wool right in my hands. My hands on her body, making her feel safe; my hands in the wool, learning what it has to teach me; my hands on the tools, converting what I have learned from Parisa herself into a yarn.

In this chapter, I want to invite you to play with the fleece with an open mind. See what it is about and how you can take advantage of any characteristics, regardless of whether they are subjective or objective, perceived as negative or positive. From that I invite you to find the superpowers of the fleece. I provide you with examples of ways to discover the fleece on your own.

From General to Unique

In chapters 1 and 2, I presented fiber types, wool types, and ways of describing wool in general. The wool you have in front of you, however, is unique. Wool description systems, breed descriptions, and breeding goals for the wool of specific breeds may be sufficient in a broader context, but here you are, a hand spinner, with every opportunity to play with and find the most interesting characteristics of this fleece.

Figure 5.1. I'm shearing the Dalapäls ewe Parisa. Two-end knitted mittens from Z-plied Värmland wool. I spun the yarn from the cut ends of teased staples to preserve as much as possible of the color variegation. (The pattern design is mine and available in the fall 2019 issue of *Spin-Off* magazine.)

To look more specifically into a fleece, I like to go to the judging charts of the Swedish fleece championships. In this, the judges use a form to grade the fleece according to a range of parameters, both hereditary and environmental: the color and the length of the staple (undercoat and outercoat) and the character of the fleece—evenness, strength, shine, softness, kemp, second cuts, and contamination. The assessments are summarized for a total point score of the fleece. Additional ways to describe the fleece are fineness of fibers and bounce. Sometimes the dominant staple type (or types) in the fleece, described in chapter 2, is mentioned, too. A few lines at the bottom of the form present an opportunity for the jury to write something more than just a quantitative evaluation. After the prize ceremony, the fleeces in the championships go to auction. As a prospective buyer to the fleeces, the few lines at the bottom of the form are sometimes the most important to me: What is the feeling the judges get when they dig their experienced hands into the fleece?

Let me give you some examples of medal-winning fleeces I have won at championship auctions: "A wool type in a category of its own. Fine fibers, soft for their length. Durable. Like a fantasy drawing had come to life." This was a Swedish Leicester/Gotland/Finull cross with 7-inch (18-centimeter) staples that got a prize for its uniqueness. The combination of softness and length intrigued me, and I spun it into a singles warp yarn for a drapey vest, and I used it together with a soft weft yarn from another medal-winning fleece.

For a bronze medalist Värmland fleece that became two pairs of two-end knitted mittens (figure 5.1) I use daily during the cold months, the jury's comment was: "Beautiful fleece with interesting color, long undercoat and soft outercoat, and a medieval touch." I was mesmerized by the color that floated between light gray, white, and honey from roots to tips. I spun them from the cut ends of teased staples to keep as much of the color variegation as possible.

A gold medalist dark brownish gray Finull/Rya cross that I blended with recycled silk into a 3-ply yarn for a warm sweater (figure 6.4) was described this way: "A very clean, soft, and pleasant spring shorn fleece. Easy to card, airy staples with mostly undercoat. An all-round fleece that is easy to work with and can be used for many things." I have a soft spot for dark gray, and, together with the softness and shine, the fleece sold itself to me the moment I saw it.

For these fleeces, the numbers and the total point from the jury of course mattered, but the subjective comments were worth even more. I have lots of fleeces that don't come from the championships; still, a judging chart is a helpful tool in finding characteristics of a fleece.

Do You Speak Wool?

Sshhhhh from rain, pitpitpit from hemlock, bloink from maple, and lastly popp of falling alder water. Alder drops make a slow music. It takes time for fine rain to traverse the scabrous rough surface of an alder leaf. The drops aren't as big as maple drops, not enough to splash, but the popp ripples the surface and sends out concentric rings. I close my eyes and listen to the voices of the rain.

—Robin Wall Kimmerer, *Braiding Sweetgrass*, 299

While listening to Robin Wall Kimmerer's experience of the voice of the rain, I wonder what language wool speaks. How can I understand what it wants to tell me about itself? My most important tool is to lean in and listen, to take in the whisper of the wool with all my senses. What do I feel, see, smell, or hear as I dig my hands into its curls? In the picking phase, my hands have experienced the whole fleece, staple by staple, and learned what it is about. They already know. All I need to do is to translate those sensations into a language that makes sense to me.

When I have a fleece in front of me in all its fleeciness, I get the chance to explore it and go on a journey in the wonders of this individual one. As in the different systems of fleece description and evaluation available, I can measure fiber length and fiber diameter, approximate outercoat to undercoat ratio, and count crimps per inch. The measurable parameters give me lots of valuable information. But they don't pull me closer to the wool. It's when I use my own words and associations that something happens; I create a bond with the fleece. When I find words to describe the wool outside the frame of metrics or even established vocabulary, the fleece becomes something personal to me. The way I dress the wool in my own words and experiences is how I will create an association with it. Opening my heart to the fleece is what allows me to bond with its soul. I need to follow the dance of the wool through exploring, preparing, and spinning to hear all the nuances in its song, understand where it wants to go, and support it in what it wants to become.

From this moment, I give myself time to lean in and play. I have a connection to the fleece; we are partners in craft, and the fleece becomes a some*one* rather than a some*thing*. This someone will gently take my hand as we travel together in the landscape that is this fleece.

A Toolbox of Words

In wine or chocolate tasting, systems of words to describe taste and experience have been developed and established. How can I describe what my hands feel in the wool when I can't find the words, when the tools in my toolbox are insufficient? My solution is to work with what I have, only in different ways than perhaps intended or expected. If I only have a hammer and no need to beat it, perhaps I can use it in an unconventional way. Perhaps I can create words that need to be created, make up metaphors or images to come closer to my experience.

Swedish is a language of compound words. By attaching existing words to each other, a longer word with a new meaning can be created, a word that might not yet exist, but no less needs to be in the world. Some of my favorite Swedish words are compound words. *Mossbelupen* ("covered with moss") and *frostnupen* ("pinched by frost") are such words, along with *fulsnygg* ("ugly in a stylish/good-looking/handsome way"). These are existing and established compound words. In my toolbox of words, though, every combination is allowed and encouraged. I can create new ones like *mineralvattenbubblig* ("sparkly, like mineral water"), *snällgrov* ("rough, but kind"), or *mosskuddig* ("cushiony, like moss"). I am the one creating this yarn. The way I describe my wool is therefore just for me. You are welcome to use these newborn words, too, though.

The Elevator Pitch

Another way of describing the wool is to imagine how it would describe itself. What would the elevator pitch of your fleece sound like? I tried this with some of my stashed fleeces. As it turned out, their pitches were expressed in poems.

The Dalapäls creamy dream expresses her features with humility and confidence, like a wise lady:

I am your Dalapäls
creamy dreamy
billowing curls.
I shine on you
like no other lock.
I give you the loftiest of loft
and the most gentle touch.
Don't think I won't provide strength,
because I will.
My wool is what you never thought
to even wish for.

A mother Rya is something special with its shine and strength. Watch her sail away:

I need no curls,
I need no loft.
I am long and strong
and will keep
this boat
afloat
as the wind fills the sails
you weave from my might.

Värmland wool can be very versatile, with many staple types and colors to play with:

I will give you all the staple types.
The long and strong
with lofty feet like ballerina skirts,
make mittens for everyone!
The fine and crimpy,
sweet like an evening breeze on your cheek.
A lacy shawl for you.
Steady, strong outercoat only,
sturdy socks perhaps?
Softy, lofty, thin tails of strength
for fulling and filling
the gaps
in mitts
In midwinter wind.

Is your fleece a poet in its elevator pitch, too? Or does it prefer a different genre of expression?

The Oohs and Aahs of a Fleece

No matter what method I use to describe my wool, I want to explore it as thoroughly as I can. I may have read up on the breed and learned about the general characteristics or about its breeding goals. Perhaps the wool from a breed has an interesting history in what textiles it has been used for traditionally. All of this is valuable and helpful, but it is the characteristics I work with here and now, in this fleece, that interest me the most. As a hand spinner, I have the chance to create a unique yarn that displays the most striking characteristics of that unique fleece.

I take advantage of the knowledge I get from the wool through my hands. With a variety of techniques, I play with the wool to get an idea of what it is about. I especially pay attention to the characteristics I find the most interesting or unique. These are the characteristics I want to focus on in a yarn. It can be a special shine, a color, strength, or versatility. These are objective parameters, but just as important is the subjective feeling I get from working with it. Through handling the fleece, my hands get a sense of how the wool behaves. All this information is useful when I pick out the main characteristics—the superpowers—of the fleece I have before me.

An individual characteristic can be found in any fleece, just as a breed description can be true for most fleeces from that breed. To me, though, it is the combination of main characteristics that makes it sparkle. This is what I want to catch and make shine in the yarn I create.

Powers Work against Me, for Me

To me, all characteristics are interesting (see chapter 2) and give me a clue to how I can process and spin it into its best yarn. The characteristics I enjoy can be candidates for those main characteristics. But others, the ones that can at first glance appear as negative, are just as intriguing. They can show me the way to process the wool more easily or present clues to how I can use the resulting yarn.

Figure 5.2. I like to make wool boards to map out the characteristics of a fleece. This one is of a Gute lamb's fleece that I blended with recycled sari silk. The board shows staples, singles, plied yarn, and three woven swatches in different degrees of fulling. The background is a whole fabric that I wove from another Gute lamb's fleece and fulled in a Swedish fulling mill.

I will give you some examples. I work with a Norsk kvit sau (Norwegian white sheep, NKS) fleece with long staples. I have plans to comb it. I notice that the tips of the staples are bundled. The Atlantic rains have pushed the lanolin out into the tips, and since the fleece has been stored for a while in my house, the lanolin has solidified. Combing the wool is cumbersome, and the wool fights me. Whenever I need to use force to process my wool, I can be sure fibers are or will be broken. To work around the struggle, I bring out my flicker (see chapter 7) and tease the tips open before I comb—just a few strokes from the outermost tip and a few centimeters in. When I comb the flicked staples, I find that enjoyable dance of the tools that my combing rhythm longs for. Exploring ways to work with the solidified tips, rather than against them, helps me decide how I can adapt my tools and the way I use them to make the process more sustainable—to me and to the wool. The yarn ended up as one of my favorite yarns ever.

Mapping It All Out

To get some structure in whatever I find from whatever methods I use, I like to take a sheet of sturdy paper, place a wool staple in the middle, and write words and scribble drawings around it, like a mind map. I can keep this paper close at hand as I go through the different steps of processing the wool and make notes as I explore the fleece. I may include objective parameters like length, crimp, and fiber type as well as subjective experiences like feeling, associations, or spontaneous thoughts of what type of yarn or textile I might create with it.

Mapping out the fleece isn't something I do in one single session. Rather, in every step of the process, from the moment I get the raw fleece, through picking, teasing, and carding or combing, I discover what the fleece is all about. The experience in my hands through the steps guides me toward how I take those steps. Playing with different tools and methods (which we will do in upcoming chapters) also gives me clues to the capacity of the fleece I work with—what is possible and what is desirable. So, consciously or not, I gather information about the fleece every time it goes through my hands.

I like to narrow the characteristics I find to three main ones—the superpowers—that I use as guides when I create my yarn. Below I give you some examples of how I have chosen the superpowers for fleeces of three breeds.

Examples

Rya wool is, as in its short elevator pitch poem above, long and strong. As my hands have played with the fleece in this example, I have chosen the main characteristics of shine, strength, and versatility. Rya does have a shine that is exceptional, and it was this, along with its strength, that was important when the breed was recreated. As it is a dual coat with usually 50 percent outercoat and 50 percent undercoat, it is also a very versatile fleece. I can separate the coats or keep them together. I can also semi-separate them into a yarn with the proportions that suit my needs. The color range of the breed also adds to the versatility.

Gestrike sheep is one of the eleven Swedish heritage and conservation breeds (see chapter 2 and figure 2.2). In this, there is not really a typical Gestrike fleece. The three characteristics I choose are thus specific to the unique fleeces I had access to for this example. Gestrike wool is a wool of contrasts; it can be very fine and crimpy at the neck and straight and rough at the sides. It is a dual coat with long and strong outercoat fibers and fine undercoat fibers. In one fleece, I found four different staple types in enough quantities to divide it into four piles for an equal number of yarn qualities. The strength of the outercoat that armors the staple allows for openness in the base, creating an airy and light fleece. Kemp is not uncommon in this breed, and I wanted to acknowledge it as something positive—the lightness it brings to the yarn and how it makes the wool

Figure 5.3. Sample yarns from four different Åsen fleeces (also figure 2.14).

easy to process. My chosen superpowers for Gestrike wool are thus rusticity, lightness, and versatility.

My last example is Åsen wool, also a heritage and conservation breed, and for this reason, breeding goals for specific characteristics are not encouraged (see chapter 2). Therefore, the wool can vary between flocks, between individuals, and over the body of a single sheep. As I have been investigating wool from Åsen sheep I have had access to staple samples from different individuals as well as entire fleeces. The fleece I worked with for this example was quite open, with mostly undercoat and just a few strands of outercoat fibers, creating staples with a triangular shape and lots of air trapped between the fibers. I experienced an interesting mix of rusticity and softness in the staples, as well as that versatility the heritage breeds usually have. Versatility and the vadmal-type staples (see chapter 2) that are quite unusual were the first two superpowers I chose. The third was kindness. Yes, with that combination of rusticity and softness—*snällgrov*—together with a soft sheen, I experienced the Åsen wool as kind.

In all three examples, versatility is one of the superpowers. This could mean a range of things, but with the background information of the breed, my playing with and exploring the fleece, and the experience in my hands, I know exactly what I mean for each of the breeds, and that is enough for me. The words in general and the combination of them for each fleece in particular are important as bearers of the information I have gathered and will keep gathering through the different steps of the process. The three words I choose for each fleece are there to remind me of what lies behind them.

PLAY: BRAINSTORM

Play with the wool to find its characteristics. Take notes of whatever you find, in whatever way you like—objective parameters, subjective thoughts, drawings. Remember, this is your fleece and your journey through it, so do this your way.

Between the official charts of describing wool for industrial purposes and the fleece's own elevator pitch poem, there are many ways of playing with and describing a fleece. Pick one, pick many. Pocket a tuft of fleece and take it for a walk, feeling the fibers in your hands as you move forward in space and nature. Perhaps you bring a question to ponder upon; perhaps you will find some clues while you walk. Perhaps the methods I have described inspire you to find your own. Take your time to play here; do what you can to make the fleece your partner in craft. Look through your notes from exercises "Greet" (chapter 1), "Chat" (chapter 2), and "Pick Your Fleece" (chapter 4) from part I, "Explore," and add what you think is important to your notes from this exercise.

When you have mapped out your fleece, take a step back and look at the result. Is there anything that truly stands out? Pick out ten characteristics that you think are important in describing your fleece. Narrow them down to five. Take your time. Then down to three. These three main characteristics are the superpowers of your fleece. What's more, it is the combination of them that makes it unique. Trust in your chosen superpowers. You know this fleece better than anyone. Your hands have been buried in this fleece numerous times and they know what it is about.

CHAPTER 6

Sampling, Swatching, and Keeping Records

The first organized physical documentation I did was for one of the first documentary spinning videos I made. I saved staples and yarn spun with different methods. With the sample yarns I wove swatches on a kitchen sponge to find the right feeling of the project. I arranged the material in a ring binder and browsed through it in the video to mark every new chapter. After that, I realized the beauty of making and keeping samples and swatches.

Often before I've acquired a fleece and certainly once I've brought it home, my hands have been in it, through washing, picking, and sorting. This gives them a deep knowledge of how the wool works (see chapter 18). When I play with different preparation and spinning methods, I convert that knowledge into a potential yarn and textile. I trust my hands to lead the way. By putting words to the physical knowledge, I can understand the wool on a more cognitive level and deepen the understanding from there.

Figure 6.1. A kitchen sponge is a perfect tool for making small weaving samples.

There are many reasons to keep records of the yarn-creating process. When I create a yarn, the exploration, play, and documentation of the process go hand in hand. My reason for keeping records is that I learn from it—making notes of what I do or experience with a fleece makes me reflect over it and go deeper into exploration. In this chapter, I want to inspire you to play with what you have; find preparation and spinning methods that work with the fleece, tools, and skills you have; and find ways of keeping records of the journey that work for you.

Through the previous chapters and exercises, you have taken notes and reflected over your fleece. In this chapter, we have a look at what we can document, when, and how.

There are endless factors you can document along the different parts of the path from fleece to textile. This is where we sort, condense, and put our notes together in a manageable format that we can come back to for reference.

The content of this chapter may seem overwhelming, but it invites you to play. Don't rush here but, instead, spend some time to find what suits you and the fleece. It'll be worth it. See the sections as tables on a buffet and the bullet points and examples as dishes on each table. Try what you like and compose your own meal. Perhaps it is a new meal every day; perhaps you skip to the dessert some days. Perhaps you aren't always that hungry.

I keep different kinds of records to different degrees and in different stages of the fleece:

- A first written documentation of the fleece and its background.
- Samples, swatches, exploration, and play to varying degrees, saved or not.
- Written documentation as I spin the yarn and of the finished textile.

Fleece Arrival: Numbers, Dates, and Figures

As soon as my fleece arrives, I do my first documentation. I like to document basic facts and thoughts on my "fiber stash" tab on Ravelry (an online community for knitters, crocheters, spinners, weavers, and dyers). These are some things I record:

- Raw fleece weight. This is my starting point. If I remember, I also weigh the fleece after it has dried from washing. Usually, I lose around one-fifth to one-third of the weight in washing. Probably more if I were to scour the wool (see chapter 3).
- Farm figures, if I know them—sheep farmer, sheep name, and breed. If possible, I ask the farmer about the sheep and the flock.
- Shearing date. Since I usually have a long queue of fleeces in my stash, I keep track of when the sheep was shorn. I try to stick to a first-in-first-out policy and spin a fleece within a year from shearing (a policy I usually fail). Since I don't scour my wool, the best before date is closer than it would be for a scoured fleece. The shearing date also tells me in what season the sheep was shorn, which gives me an indication of the amount of lanolin, dirt, and vegetable matter (see chapter 2).
- Scorecard, if the fleece comes from a competition.
- Photo of the fleece. This makes the fleece easy to find on my Ravelry page. It is a great reference to have once the wool has all been spun. Also, my sense of organizing can't stand a project page without a photo.
- If I have ideas of what to create from the fleece, I take notes of that, too.

Figure 6.2. A sheep owner keeps track of her fleeces in basket drawers.

Fleece Exploration: Play, Sample, and Swatch

Through the picking (see chapter 4), I may have had thoughts about tools and techniques for wool preparation and spinning, textile techniques and textile projects—anything I see, feel, and think of in this phase is worth documenting.

Since I usually store and queue my fleece after I have picked and perhaps sorted it into categories, it is important to me to make this documentation. When I get to that fleece in the queue, I am very grateful for the notes and reflections I made up to a year earlier.

Translate

In chapter 5, we looked at the characteristics of the fleece and narrowed them down to three. The superpowers themselves, the explorations and choices that led to those characteristics, and the combination of them are our working material. We need to translate the superpowers of the wool into a yarn and, eventually, a textile. We do that with the help of our knowledge of wool and fiber in general, this fleece in particular, and the tools and techniques at our disposal.

If the fibers are fine and soft, I may want to spin a yarn for a next-to-skin textile. Fine fibers usually come from the undercoat that is meant to keep the sheep warm. I therefore may choose to prepare and spin the wool with techniques that allow air between the fibers. Carding rolags and spinning a woolen 2- or 3-ply yarn for knitting may come to mind. I could spin a rougher fleece in this manner, too, perhaps for outerwear.

With long and strong fibers, I may want to spin a yarn for strength and durability. The fleece may have mainly outercoat fibers that are designed to keep the sheep dry. I could choose techniques that align the fibers—perhaps I comb the wool and spin it worsted (see chapter 10) for a singles or 2-ply warp. I may spin a fleece where I want to enhance the shine in this way, too, perhaps with embroidery as the end product.

If the fleece has different colors that I want to divide it into, it's possible that the colors have different characteristics. In my experience, white wool is generally finer than colored wool. I may use different techniques for different colors for that reason.

I spun the Dalapäls "creamy dream" in the elevator pitch from chapter 5 into a strong and shiny Z-plied yarn for two-end knitting. To get an even distribution of outercoat and undercoat, I chose to tease lock by lock with a flicker and then by hand into an accordion burrito preparation (see chapter 7). The preparation was also easy to travel with, and I spun a lot of the yarn on a supported spindle on train rides. The yarn ended up in a pair of two-end knitted jacket sleeves.

I spun the Åsen wool in chapter 5—versatile, kind, and with vadmal-type staples—woolen on a suspended spindle. The woolen technique (see chapter 10) gave loft and warmth, while the suspended spindle gave it some strength from its weight. I nalbound a pair of mittens and waulked them into shape and size. I brushed the inside and ended up with mittens that were surprisingly strong and soft. You can find more examples in chapter 5.

Separate or Not?

Sometimes I spin different yarns from different parts of a fleece. Perhaps I have decided at the picking stage to divide the fleece into categories (see chapter 4). I can plan and create the yarns for different projects. I could also spin different yarns for different parts of the same project. This could be two different sock yarns—one strong for the foot and calf and one even stronger for toes and heels. Another option is different yarns for a sweater with a softer yarn for the neck, a medium yarn for the body, and a stronger yarn for the elbows.

With an Icelandic fleece, I separated the tog and the thel (outercoat and undercoat in Icelandic wool) by hand. I did this, first, because I enjoyed the feeling of it, but also to be able to sort the fiber types according to color, too. I ended up with light gray, almost white tog, as well as an anthracite color. The color variation in the thel was smaller, but I did get a medium gray and a light gray. I combed the tog and spun worsted singles from two colors for a striped warp and carded the lightest thel for a woolen singles weft.

A similar project was a cross between Swedish Härjedal and Åsen sheep. It had a multitude of brown colors over the fleece as well as between outercoat and undercoat. I sorted the staples into four color piles and separated the coats. I ended up with strong and shiny weaving yarn in four different colors and soft and airy knitting yarn in another four colors. I used the weaving yarn for a striped belt bag and the knitting yarn for a gradient hat.

Figure 6.3. The hat is knit with yarn I spun from the undercoat of a multicolor fleece that I divided into four color categories. The brown stripes in the bag come from the outercoat of the same fleece. The dyed stripes are the outercoat of a Klövsjö fleece. (The hat pattern is by Fiber Tales; the bag is based on a sewing pattern from Merchant & Mills.)

Having a fleece with different qualities doesn't mean I have to separate it, though. I can choose to blend it to create yarn that combines strength, shine, and loft for a medium yarn. One example is a Rya fleece, a dual coat with long and very shiny and strong outercoat and softer and lofty undercoat. I decided to keep it all together in a 2-ply yarn for rya knots in a woven bench pad. The outercoat provided strength and shine, and the undercoat filled out the empty spaces for that abundant texture.

Sometimes I like to semi-separate a dual coat. Perhaps I want a soft yarn with a little strength. I might remove some but not all of the outercoat for this purpose or separate fully and then add some outercoat back. A dual coat can in this way give an endless array of opportunities.

Blend

Another way to play and explore is to blend. The fleece of a cross can have that perfect blend that gives me the quality I am looking for, but I can also choose to blend different fleeces or fibers to get what I want. For a sock yarn, I blended a Rya fleece with adult mohair. The Rya outercoat gave strength and shine, the undercoat contributed softness and bulk, and the mohair boosted strength and shine. One of my first handspun projects was a Finull yarn that I had blended with mulberry silk at the teasing stage (see the sweater I'm wearing in figure I.1). The silk added strength and softness to the yarn. Also, the different fibers took dye differently and provided a gentle color variegation.

Unexpected Choices

When I explore and plan a project, the characteristics are the focus of my attention. Sometimes, though, I make unexpected choices. These can come from playing with the wool and finding an exciting technique or result, or through a conscious choice based on a feeling or the context I will work in. The Dalapäls creamy dream (above and chapter 5) was such a choice. I planned the preparation technique partly based on my wish to be able to spin it on the go. Chance led me to spin a whole year's growth of Jämtland fleece from the fold from flicked staples. Combing with a combing station would be a lot faster, but I realized that spinning from the fold gave an increased loft in the yarn and ease in the spinning.

Opposites

When I get a feeling for what I want to do with a fleece, I play and swatch with techniques to come close to what I imagined. Sometimes, though, I also dare myself to do the opposite—the unexpected, or something I may be afraid of trying—just to see what happens. I can stick to preparing and spinning a fleece in a certain way, but it is not until I compare it to another method that I can see and understand the best result.

If I envision the wool as a medium yarn, I may spin a fine, medium, and bulky yarn just to see and feel what I can get (figure 6.4). Perhaps the samples will confirm why the medium is the best choice, or I will realize that the bulky yarn works better and why. Once I spun a Gute yarn that I had blended with recycled sari silk for a weaving project. Just for fun I tried fulling a woven swatch to see what happened to the silk (I was convinced it wouldn't felt) and was surprised by the blurred little specks of color felted deeply into the structure. I wove another swatch but fulled it a little less. I realized I had found the perfect texture—a lightly fulled fabric with a subtle drape that I hadn't even imagined possible (figure 5.2). Trying opposites is a great way to learn how the wool behaves.

Challenges

In the wool characteristics section in chapter 2, I describe the characteristics in neutral terms, and I do my best not to mark them as negative or positive. I believe we as spinners need to do our best for the fleece we have instead of wishing it were something it is not. If I encounter a characteristic as a challenge, I want to explore it to find a way to process and spin it that takes advantage of the characteristic. If there are characteristics that are difficult to work with, I want to find tools and techniques that make it easier on both me and the fibers. The Norsk kvit sau (NKS) fleece in chapter 2 is one such example.

Putting It All Together

With all the playing above I can make myself quite a collection of samples and swatches. I need to make sense of these and use them as a guide to my yarn, together with thoughts on main characteristics, tools, spinning techniques, and textile techniques.

Wool Boards

I don't document physical samples for all the fleeces I work with—far from it. I never regret the work when I do, though. When I do document samples and swatches, I keep records of my tryouts and, especially, the methods I decide to implement later. I may take notes in a notebook or on Ravelry. I may make what I call a wool board with samples from staple to yarn and describe the main characteristics and the methods I have used. One of my wool boards is usually a piece of stiff paper where I have attached samples and swatches and condensed my most important reflections. At the moment I have less than ten elaborate wool boards and a number of labeled and unlabeled swatches in my box of records. My wool boards can contain

- typical staples of the fleece;
- samples of singles, plied yarn, or tryouts;
- swatches of textile techniques, like knitting, weaving, crochet, and nalbinding, etc.; and
- fulled woven swatches.

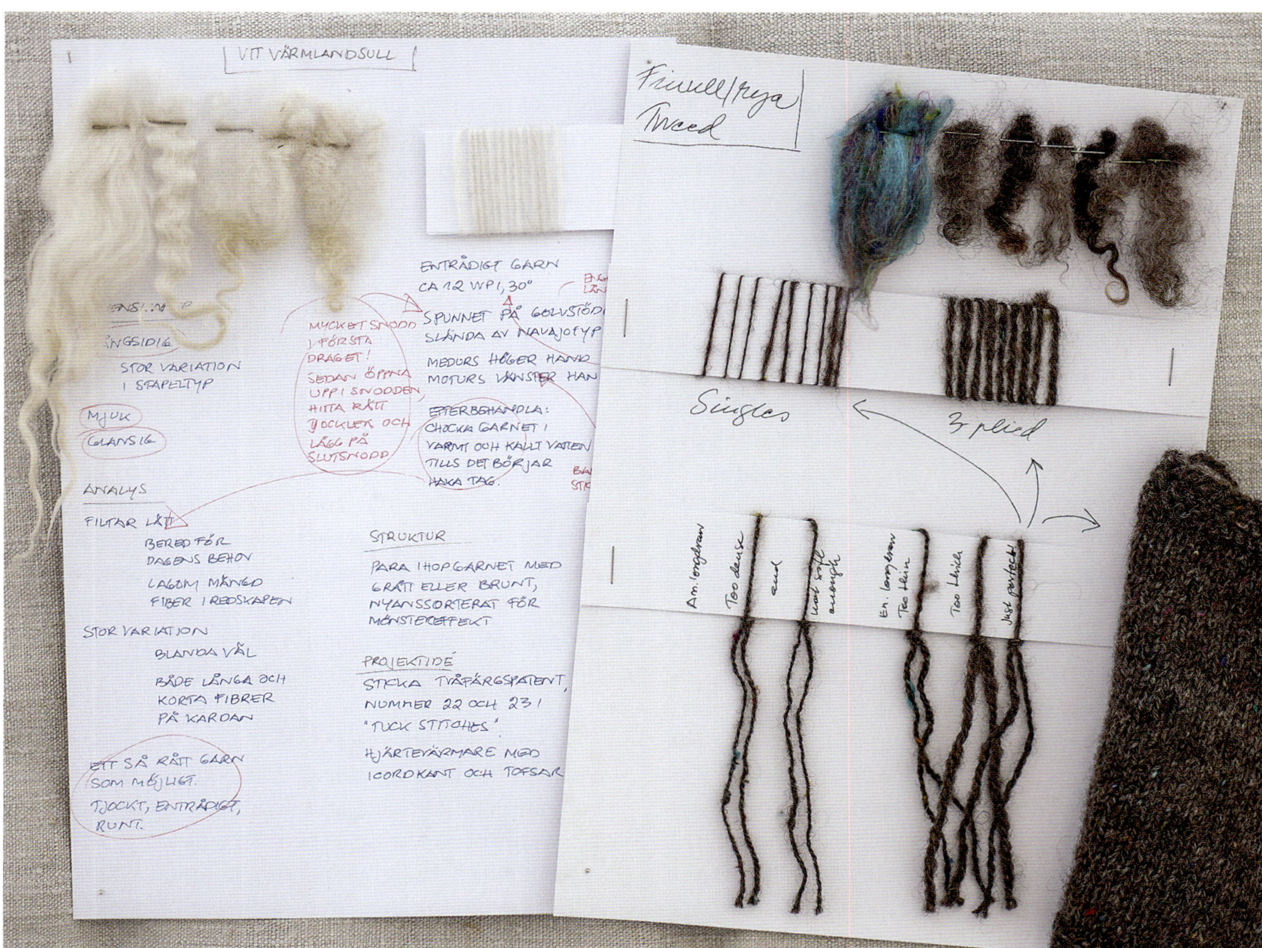

Figure 6.4. Some examples of wool boards and what I like to include in them—staples, singles, plied yarns, and swatches—plus notes of my thoughts about them. The white Värmland wool later became a thick singles yarn for a shawl in tuck stitches. The dark gray Rya/Finull cross became a 3-ply yarn for a sweater that I knit as a prototype for a sweater pattern I created for another fleece from the same flock.

I also add notes of my chosen superpowers, other important characteristics, planned tools and techniques, things to be observant of, like any challenges (see above). I may also write something about why I have chosen a certain technique or tool. Sometimes I write questions to myself as a reminder or as something to explore further.

"Do I really need to swatch?" you may ask, just as I have, many times. And too often I think, "Nah . . ." Swatching, however, is a way to see how a yarn looks and behaves in a fabric or in different techniques of that fabric. It is also a way to learn how the yarn feels as I create the fabric. The process doesn't end when the fleece basket is empty–a yarn basket is now full, and I continue the yarn's journey to a fabric. I want the creation of the fabric to be just as joyful as the creation of the yarn.

When I make a wool board, I use it for reference in the yarn I'm spinning and in other projects. Even if I never look at a wool board again, I have learned just by having reflected and put my thoughts on paper. I have boiled down my thoughts and come to a better understanding of the characteristics of the wool and a clearer view of what I want to create.

A System (or Not) to Fit Your Needs

This is my way of keeping records. I don't always do it the same way, and I don't always document the same things. To get a picture of how other spinners keep their records, I put together a survey and sent it to my readers and students. From the many replies I received, it seems like every spinner has their own way and extent of keeping records. Some keep a very consistent system; others adapt their record keeping to the wool or the situation. Some do a minimum, and some keep no documentation at all.

Many of the respondents expressed a curiosity about other spinners' ways of keeping records and a striving to find a way that suits them. Between the lines I interpreted an ongoing search for one system that works all the time.

One interesting thing I noticed was that there seems to be an expectation of record keeping. Some respondents expressed guilt about not documenting enough or of documenting the wrong way. It seems that spinning teachers talk about documentation as a must, that there is a "talk on the town" about record keeping. "People seem to think you have to, so there's always conversation about it," one respondent writes. "I am sure I would document more if I had the knowledge or a book like the one you are writing, I think. I feel a little lost in this." (My translation from Swedish.)

When I ask about what the respondents use their documentation for, many of those who have expressed insecurities about record keeping reply that they don't know. "Not sure yet, but every YouTuber in spinning told me to keep records." Others answer that they keep records for consistency and for the possibility to match yarns, to experiment, to learn, and to remember if they have a longer break in the project. Yet some keep their records to be able to see their progress and for nostalgic reasons. "Sometimes I just like to page through my records for inspiration or to see my own progress, like a photo album but for my yarn children."

In some responses, I see a sense of a dichotomy between keeping records and spinning for the joy and the process, with a concern that the documentation will take something away from the spinning process. "The pleasure of spinning is the most important reason I do it. Keeping documentation is not pleasant for me," is one example of this kind of reply. One respondent takes this dichotomy to a deeper level: "I have a divided feeling toward documentation. I can enjoy pretty labels and love to be organized. But I also struggle against the feeling that what isn't documented isn't worth anything. So I can wish that I documented more and that I documented less? I like to document, but perhaps I wish I didn't? I wish that not everything needed to be a documented experience to fall back on but just something pleasant that I engaged in for a while and that later was allowed to be forgotten" (my translation from Swedish).

Yarn Journey: Notes of Finished Yarn and Textile

My record keeping continues through creating the yarn and textile. I document my choices on my Ravelry "handspun" and "project" tabs. On my handspun tab I may note

- tools and techniques for preparation;
- spinning tool;
- spinning technique, including settings like wheel drive ratio, number of treadles for adding twist and for plying, number of plies, and twist direction;
- length and weight per skein;
- grist—finished yarn length per weight unit, in my case meters per kilogram;
- how much of the fleece I have used for the spinning project; and
- other reminders to myself.

I also add photos of the finished yarn. I could add a point about twist angle (the angle of the fibers in the yarn, indicating the degree of twist) in singles and plies, but honestly, I never do. I seem to go by rhythm and intuition to find the right amount of twist and an even twist throughout the project (see chapter 10).

I always make a yarn sample when I start spinning my planned project and hook it over one of the maidens of my spinning wheel. Sometimes this is my only physical documentation.

The grist calculation (yarn thickness, usually expressed in length per weight) is something I do for every single skein. With this number I have something to aim for in pursuit of an even yarn quality over all the skeins. As soon as the grist in one skein deviates too much from the grist of the previous ones, I know I need to adjust something. When I use the yarn in a project, I can make sure I place the deviating yarns smartly, though. For a sweater I may place a higher grist skein (finer yarn) at the neck rather than at the hip for the best fit.

On my Ravelry project tab I may make notes of how the yarn behaves in the textile I create. In my first weave with a handspun singles warp, I realized that the energy of the singles twisted the warp as I rolled it onto the warp beam of my rigid heddle loom. I made a note to never let the warp go as I rolled it on, to avoid tangles in the weaving. I also made a note of winding my singles onto tennis balls a few weeks before warping, to tame some of that energy.

Using the Documentation

Throughout this chapter, I have given examples of the documentation I make and how I use it for particular projects. But I also use my documentation for a broader purpose.

One type of documentation that I benefit from on a larger scale is the raw fleece weight connected to the finished yarn weight. With these numbers, I can calculate the raw fleece to finished yarn yield for each fleece I spin. Having made this calculation through nearly all my fleeces, I now know that my usual yield is around 55 percent. If I aim toward one grist for all the skeins of one fleece, I can roughly estimate a final meterage for the whole fleece if I calculate from that 55 percent yield.

If I have spun a yarn that I particularly liked, I can go back to my notes and use them for another project or use them as a starting point to develop another technique or quality. One example is a 3-ply woolen yarn I spun for a sweater (figure 6.4). I deliberately designed the sweater as a beta version for a more elaborate design. The fleece came from a dark gray Finull/Rya cross lamb, and I asked the sheep farmer for a similar fleece, in white, for the later version. I used my beta swatches and thoughts for the new yarn and fine-tuned it for the new design that eventually turned out as the Selma Margau sweater (figure 6.5). The pattern is available in the Summer 2020 issue of *Spin-Off* magazine. As I had yarn left from both the beta and the final versions, I spun a third yarn from a Finull/East Friesian cross that I divided into three shades of gray and created a third sweater with a gradient of the five colors from the three fleeces (figure BM.2, p. 173). Without the notes from the first and second projects, it would have been difficult to create such a match.

Figure 6.5. A 3-ply Rya/Finull cross with recycled sari silk. I later made the Selma Margau sweater pattern from this, with the foundation of the prototype on my wool board (figure 6.4).

PLAY: MAP

Now, gather your notes and exercises from chapters 1–5 and put together the ones you find the most important for creating your yarn. You can do this at any scale—just a few notes and samples or a larger fleece study. Play with your fleece to discover the yarn(s) you want to spin from it. Gather your tryouts and favorite samples and make a few swatches in one or more textile techniques. Create your documentation with inspiration from the examples in the chapter and from other spinners. Reflect on why you want to (or why you choose not to) keep records and shape your documentation from there. This is for you, in your way and for the purpose you decide. If nothing else, make it as a guide for yourself and a tribute to the sheep that gave you the wool.

CHAPTER 7

Add Air: Tease Your Wool

Through picking, the wool has been slightly opened and is now easier to handle. At the teasing stage, I will continue to add air into the wool to create an even distribution of the fibers. Teasing destroys the staple shape and readies the wool mass for structuring and shaping in the carding step. I generally do not tease before combing.

When I look in both Mabel Ross's *The Encyclopedia of Handspinning* and Enid Anderson's *The Spinner's Encyclopedia*, teasing is referred to as opening up, staple by staple, by hand into a film-like distribution. Neither of the books describes a picking stage. I interpret the definitions in the encyclopedias as somewhat of a fusion between my description of picking and teasing.

In this chapter, I want you to understand why we tease the wool and how this seemingly time-consuming step can actually save both time and energy, deepen our experience, and maintain quality and yield in the upcoming steps of the process. I serve you with teasing techniques with different tools, depending on what tools you have, what result you are striving for, and what your wool is like.

Figure 7.1. Teasing wool before carding is a gift to myself.

Why Tease?

When I tease, I prepare my wool for carding. The time and effort I invest at the teasing stage is time and strain saved and quality enhanced in carding. I have talked to many students and readers who have abandoned hand carding because they find it tedious, frustrating, and straining. At the same time, other students who have carded for decades and never teased have found a new joy of carding and created higher quality rolags after having included teasing in their practice.

Another way of understanding the importance of teasing is to think of woodwork. To smooth a rough piece of wood, you wouldn't go straight to the finest sandpaper. It would require a lot more work and strain on your body and a lot more waste of material. Instead, you would start with a rough sandpaper and go down to the finest in a couple of steps. Just as the picking (rough) prepares for teasing, teasing (medium) prepares for carding (fine sandpaper), and makes each step easier on your body as well as on the fibers. In figures 2.17 and II.1 you can see the raw wool, picked staples, teased wool, and carded wool. Inviting air in between the fibers also helps remove vegetable matter, directly and indirectly. Some pieces of vegetable matter fall out as the staples open; others will be more available to pick out manually.

Teasing can be the only step of preparation. In the Andes, a tradition is to tease the washed wool by hand in the evening to prepare it for spinning the next day.

On a Rock in the Forest: Teasing with Your Hands

First off is a very reliable teasing tool: your hands. Teasing with your hands takes a lot of time, but it is also the cheapest and simplest way to tease. You can do it sitting on a rock in the forest and get really close to the fibers.

To tease my wool, I hold a staple lengthwise in one hand. With the other hand, I pull just a few fibers sideways, perpendicular to the direction of the staple, and repeat until the whole staple is teased. The fibers leave the staple in a bow.

As the fibers go through my hands, I have the opportunity to get to know them—their length, structure, crimp, elasticity, and glide. Even if I don't always actively take note of these characteristics, my hands do. The next time I sit with my fleece and tease, the hands remember and adapt to the properties of the wool (see chapter 18).

Figure 7.2. When I tease with my hands, I hold the staple parallel to my thumb and pull just a few fibers from the center of the staple and outward in a bow.

Teasing by hand has its advantages: I don't need any tools, it is a cheap option, and I can tease anywhere in limited space. I also find it very satisfying and calming to just spend time with the staples. The time it takes can be a disadvantage. I also find that hand teasing may not always remove fibers that are too short to spin.

One Staple after Another: Teasing with Cards and Flicker

If you are planning to card your wool after teasing, you may already have a pair of hand cards. You can use one of them to tease your wool.

Place the card in your lap, teeth facing up. Place the cut end of a staple on the top side of the card. If the cut end is tangled or felted, you can tease it apart sideways slightly with your hands first. To keep the staple in place, gently hold a hand on top of the staple. With the other hand, pull the tip end straight out from the card, wiggling it if the fibers are reluctant to leave, or initially tap the staple against the teeth. Make sure you hold the staple quite close to the cut end. Otherwise, there is a risk that too many fibers stay in the card. The cut end will glide through the teeth and separate. Shorter fibers may stick to the card. If they are about 2 inches (5 centimeters) or longer, pull them straight out, too. Fibers shorter than

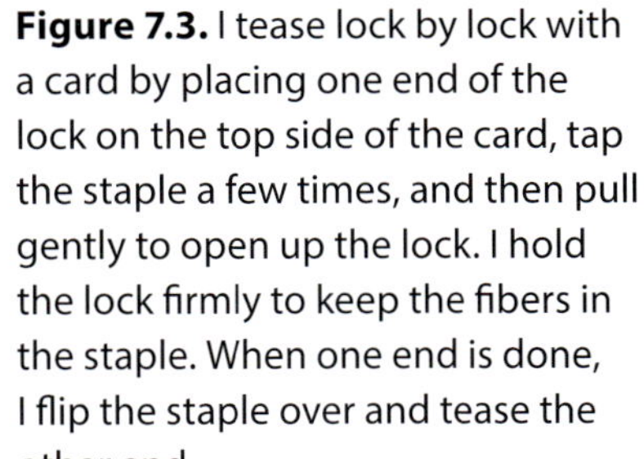

Figure 7.3. I tease lock by lock with a card by placing one end of the lock on the top side of the card, tap the staple a few times, and then pull gently to open up the lock. I hold the lock firmly to keep the fibers in the staple. When one end is done, I flip the staple over and tease the other end.

Figure 7.4. I work the flicker similar to teasing with a hand card. I place a piece of leather underneath the flicker to protect my clothes.

that will be a challenge to incorporate in the spinning twist and may end up as nepps, so I would lift them up from the card and put them on the waste pile. Flip the staple when the cut end is evenly teased and tease the tip end the same way.

A flicker is a small card designed for teasing. The carding pad can be horizontal or vertical. There are also flicker blocks on the market that can be mounted onto a table (see appendix 2). The flicker technique is the same as teasing with a hand card, only a bit slimmed down.

Hold the flicker, teeth down, in your lap on top of the cut end of a staple. I use a piece of rough cloth or leather underneath the flicker to protect my clothes. Tease the cut end open by moving the flicker toward your body. Flip the staple to do the same on the tip end.

Teasing with a card or flicker is more efficient than teasing with your hands. I like to use this method with wool that is very fine. Swedish Finull wool, for example (see chapter 2), has fine fibers that can get brittle in the tip ends. During the carding, these tips can break and leave nepps. If I tease the staples one by one with a card or flicker, any brittle ends tend to break in the teasing and stay in the flicker instead of breaking in the carding. I also like to use a flicker or a card when the staples have a lot of kemp (or too short fibers in general) in the cut ends. A flicker or a card can do a good job in removing some of the kemp (figures 1.2 and 2.17). If I plan to spin the staple in its teased form, either from end to end or folded over my finger, teasing with a flicker would be my first choice.

Efficiency: Combs

The most time-efficient way to tease is with combs. They can be charged with several staples and tease a larger amount in one go. I have several pairs of mini combs that I use for myself and for my courses. I use both mini combs and larger combs with a combing station for both teasing and combing. While teasing with mini combs is a process I enjoy, teasing with a combing station feels more like a production activity. I like to clamp the combing station to an adjustable desk and comb standing. The description and images below are for teasing with mini combs. For teasing and combing with a combing station, see chapter 8.

I hold one mini comb in each hand. For ease of explanation, I call the comb in my dominant hand the *active comb* and the other comb the *stationary comb*. In the same manner, I call the fiber ends in the tines the *stationary end*, and the fiber ends in the air the *active end*. I charge the tines of the stationary comb with a handful of picked staples. When I tease wool with combs, I don't need to bother about any direction of cut ends or tip ends. It's important, though, that as little as possible of the wool peeks out on the handle side of the comb, and that I charge to only one-quarter of the height of the teeth. Any more than that will be too heavy to work with and unevenly teased. The active ends now point out from the front of the comb.

I hold the combs perpendicular to each other, the stationary comb vertical (teeth pointing up) and the active comb horizontal (teeth pointing away from me). In this way, the teeth of the combs form a grid when held facing each other. The combs will be held in this perpendicular position all through the combing. The only thing that is moving is the active comb. I will move my active comb in two circles: one horizontal and one vertical. The orientation of the circle is the only thing changing.

Figure 7.5. Teasing with mini combs: I charge the comb (with no consideration of staple direction) with a handful of staples and make sure as little as possible is showing on the handle side of the comb. I charge to about one-quarter of the height of the tines. (Sweater pattern by Quince & Co., knit in commercial linen yarn.)

Figure 7.6. Teasing and combing with mini combs: I hold the combs perpendicular to each other, one with the tines up and the other with the tines out. With the active comb, I make a horizontal circle and catch the tips of the staples when I move into the stationary comb.

Figure 7.7. Teasing and combing with mini combs: When I have caught the wool with the active comb, I make a vertical circle with the same comb to transfer the wool back to the stationary comb.

Figure 7.8. Teasing with mini combs: When the wool is airy and open, I pull it straight out.

I start the horizontal circular movement with my active comb, catching the outermost parts of the active staple ends on the stationary comb. As I keep going with my horizontal circle, I move closer toward the teeth of the stationary comb as the fibers detangle. When as much as possible of the fibers have transferred to the active comb I switch to the vertical circle. In this part I transfer the new active ends of the fibers back to the stationary comb. When the active comb is as empty as it can be, I switch back again to the horizontal circle. Any fibers that stick to the emptying comb are either tangled or too short to join in the twist of the yarn.

When the fibers are evenly teased, perhaps after three to six turns, it is time to remove them from whatever comb they are on. The teased fibers stick out of the comb perpendicular to the teeth. Any vegetable matter, short fibers, nepps, second cuts, or tangles are stuck between the teeth. The temptation is to pull the fibers up in the direction of the teeth. This will leave me with both teased fibers and waste. Instead, I pull the fibers straight out from the teeth. The teased fibers end up in my hand and the waste will remain in the comb. I remove the waste and go on to tease the next round of staples.

No matter which method you try, look in your lap and on the floor. Do you see vegetable matter? If you do, this is stuff that won't bother you in the upcoming steps. Congratulate yourself on having further tidied your wool.

Blending

If I want to blend my wool with another fleece, color, or material, teasing is usually the stage when I do this—preferably with combs, but the other methods work well, too. My favorite way of blending is with stationary combs. It's efficient and I find it easier to create an even blend. When I charge the comb, I add the fibers I want to blend and start teasing. The fibers will blend evenly, and I will have a unique mix to work with.

Combinations

Recently, I got a variegated fleece of an Icelandic sheep in my lap. I had read an article about spinning a lopi-style yarn and decided to try it for myself. I wanted to spin the yarn with as little manipulation of the wool as possible to create a raw look and feeling. I decided to spin in the grease (see chapter 3) from the cut ends of teased locks.

The staples were cone shaped, airy, and very long, in some cases 9 to 12 inches (20 to 30 centimeters). I played around with different teasing methods and found a combination of methods that worked very well. I teased each staple individually with a flicker in the way described above. After that I opened up the teased staple sideways with my hands like an accordion. The double-teased staple was now stretched out into one long and thin layer. I finished by rolling each stretched-out staple into a loose burrito. This way the fibers were still aligned in the same orientation as in

Figure 7.9. Sometimes I want to disturb the staple formation as little as possible and tease with a combination of flicker and hands. I open the staple with the flicker, spread it between my hands into an accordion, and roll one or more of these stretches together into a burrito shape. I spin these from the cut ends.

Figure 7.10. Willowing wool is an old technique for bringing air into the wool, removing vegetable matter and blending different wool qualities or colors. (Sweater pattern by ANKESTRICK, knit in commercial linen yarn.)

their stapled form, but airily and evenly distributed. With this shape, it was easy to hold the burrito preparation in my hand and spin from the cut ends of the loose preparation.

Willowing

Willowing is an alternative mechanical way to open up a fleece instead of picking and teasing it or in addition to either of these techniques. Willowing means to beat the fleece with flexible sticks, usually willow. Air will be invited in between the fibers, making the wool puff up. After a while, the loosened staples will start flying in the air above the pile. Vegetable matter flies out from between the fibers and lands on the ground.

Beating the wool with sticks used to be an occupation in European medieval times, called *Ullpiskare* in Swedish ("wool whipper") and *Wollschläger* in German ("wool beater"). *Wollschläger* is still a German surname. The wool beater whipped the wool to open it up and sometimes also blended different qualities or colors together to get the desired result. The wool beaters also traded in wool. The finished blends would then be sold to wool manufacturers.

When I first encountered willowing in a spinning course in Sweden, I was told that the staples flying above the wool pile were traditionally called angels. The higher the angels flew, the better the wool was beaten.

Willowing wool is quite fun, sitting on the ground, beating the soft wool and watching the angels fly above like soap bubbles. There is a kind of childlike innocence about it.

Joyful

Let's create a scenario. Let's say I didn't tease my wool; perhaps I didn't even pick it. The wool would be quite dense, with lots of vegetable matter and unwanted bits. Carding would be a lot more straining for my body. Fibers would break, and there would be a lot more waste. My rolags would be less even in both fiber distribution and shape. With the extra force I would have to use to card the dense staples I would probably not get to know the wool like I would with ease in my movements. I dare say the carding wouldn't be joyful. The yarn I would spin from rolags carded straight from staples would be uneven and bumpy, perhaps it would break easily. The process would be more of a struggle.

Every part of the wool preparation is a joy to me. I want it to be, and so I make sure it is, through dividing it into smaller parts. Compare it again with the differently grained sandpapers. Furthermore, every part of the preparation paves the way for joy in the next step. The wool I tease has already been prepared in the picking stage. The staples are separated and airy; second cuts, felted parts, and some of the vegetable matter are gone. Perhaps the staples are sorted into categories. In the picking, I have given myself the gift of airy staples ready to be teased. In the teasing, I will prepare the staples further for the carding stage and give my carding self another gift.

Transformation

After teasing, the fleece has transformed fully, from staples once bundled together to a homogeneous mass of evenly distributed fibers. The sheep wouldn't recognize its own coat. Moreover, every fiber has gone through my hands in every step, and every fiber has been inspected, tested, and moved through my attentive fingers. The characteristics of the fibers tremble as memories on my skin. Regardless of what tool I used, I have provoked the fibers in different ways. In their reaction they have responded straight into my hands with information, pieces of the spinning puzzle. Every brush with card or flicker has generated a countermovement from the fibers. Every time my fingers have teased out fibers from a single staple they have felt with what speed, friction, glide, and elasticity the fibers have left the staple. In every pass with the combs, my hands have adapted to the characteristics of the fibers as they have parted to give way to air between them.

PLAY: OPEN

Now, tease your wool. Try as many ways as you have the tools for. Tease different kinds of wool. See what works for you and what works for the wool. As you tease the wool, take notes of what you feel—the glide of the fibers against each other, the resistance from the wool as you work with your hands or tools, the elasticity, and the bounce. Watch the vegetable matter release and fall into your lap or onto the floor. Reflect on the differences and similarities in using different tools and working with different fleeces.

Figure 8.1. Carding by hand is a joyful activity, provided the wool has been opened and teased first.

CHAPTER 8

Prepare: Shape the Mass and Structure the Fibers with Cards and Combs

In this chapter, I will share my methods of carding and combing, and why I use specific techniques. I also invite you to play with the fiber you have before you, to see what the fibers can do in different kinds of preparations and how we as hand spinners have a spectrum of choices when we work from a fleece.

I can choose to align the fibers from picked staples or structure the teased wool and shape the mass. It is perfectly doable to spin straight off teased wool or even picked staples. With a combed or carded preparation, however, the wool is prepared on a deeper level, into consistent units that offer a more even flow and a higher quality yarn.

Preparing the fleece is part of the process of creating yarn. To me it is important to make all the steps from fleece to yarn and finished textile joyful. If it is not, I need to make it so by choosing tools and techniques carefully. In appendix 2, I have listed tools I use and qualities I look for.

You won't read anything about drum carding in this book. This is simply because I don't use a drum carder. I prepare all my fleece with hand tools because I enjoy it, and I get to know the wool with my hands in the fleece. I am also certain that working by hand gives me the best results.

Carding

Why I Card

I card to get a woolen preparation—an airy, structured, and shaped preparation. I arrange this in equal units: cylindrical rolls called rolags. The fibers may

Figure 8.2. Rolags are carded batts that have been rolled into a cylinder.

be somewhat jumbled in the preparation, but with an even distribution and lots of air between them. With evenly carded rolags, I have a good chance of spinning an even and airy yarn with a smooth flow.

To me, carding is about structuring the fibers and shaping the mass. Therefore, I card only teased fibers. Carding unteased staples will create lots of strain in my body and on the fibers. The process will be anything but joyful (see chapter 7). I believe that one reason spinners abandon their hand cards, often before they even get the hang of the method is that they card unteased wool, find it strenuous, and never get to feel that carding joy.

I usually card to spin a woolen yarn (see chapter 10), but it is possible to spin carded wool with other techniques, too.

What I Card

Short fibers, perhaps 2–4 inches (5–10 centimeters), are perfect to card. With only longer fibers the strands will double up in the rolags and create tangles. A combination of short and long fibers, however, is a good choice for carding, preferably with a scale of lengths. The short fibers will provide the bulk of the rolags and the longer will armor the construction, while the medium fibers will marry them together. This means that fibers longer than 8 inches (20 centimeters) can work in a carded preparation, as long as they are accompanied by a reasonably even scale of fibers down to around 2 inches (5 centimeters) (figure 2.16).

How I Card

There are many ways to card wool, and every carder has their preferred method. I have carded different ways through the years, but around 2017 I watched a video with four spinners who all had their personal carding style, and reasons for every detail of it. I watched the video several times and baked my own carding style from the ingredients the spinners provided. By sharing my technique and why I do each step, I invite you to do the same.

I use the term *active card* for the card that is moving and *stationary card* for the card that is not. I call the edge of the card with the handle the *handle side*, and the opposite long end the *top side*.

Start: Dress and Frame

In every step of the carding process, I work with light movements. The dressing of the card is the first step

Figure 8.3. I dress the carding cloth by gently stroking teased wool onto it. Wool that doesn't stick is superfluous, and I leave it for the next rolag.

Figure 8.4. I leave a frame around the wool empty when I dress the card.

and the foundation of the shape and quality of the rolag. To dress the card, I hold my stationary card in my nondominant hand, teeth up, and let the card rest on my thigh, long side parallel to the length of my thigh. I take a light handful of teased wool (see chapter 7) in my dominant hand and gently stroke the stationary carding cloth with the wool, starting about an inch (a few centimeters) from the top of the handle side of the cloth and dress across the teeth to the top side. As I do this, I gently pull the fibers, such that the fibers stick to the teeth (but are not buried in them) at the place of onset, and are lightly stretched to the top side. I keep doing this across the carding cloth, leaving about an inch (a few centimeters) empty at the short ends, too. I only charge what sticks to the carding cloth, meaning that any wool that doesn't stick stays off. Through the years, I have carded thousands of rolags. By now my hands know exactly how much wool is enough for one rolag. Eventually yours will, too.

I now have teased wool lightly dressed on the card, with a frame of the carding cloth around the wool empty. As I card, the wool will fluff up and out. Had I charged the wool all the way to the edges, it would fluff outside the carding cloth, and the wool would be unevenly carded.

Middle: Structure and Transfer

Now, for the carding part, I take the active card in my dominant hand, teeth down. I hold it with my pinkie, ring finger, and middle finger around the handle and my thumb and forefinger straight in a *V* shape over the back of the paddle to stabilize it. This is where the design of the cards can decide how you card. Some cards have curved paddles; some have flat ones. I like to follow the shape of the paddle as I card—a rocking motion for curved cards and a straight, horizontal motion for flat cards. With the active card, I lightly stroke the wool on the stationary card as follows:

- with flat cards: from the handle side of the stationary card, straight over the cloth to the top side (moving each stroke away from the handle side of my stationary card [figure 8.5]).
- with curved cards: from the top side of the stationary card in rocking motions three to five times up to the handle side (thus moving the rocking motion toward the handle side of my stationary card [figure 8.6]).

Again, the key word in carding is lightness. The teeth of the carding cloths are bent toward the handle side, such that the teeth of the stationary and active cards point away from each other when I card. The task of the cards is to stretch the fibers between them and separate the fibers to invite air. This means that, rather than pushing the fibers into the teeth, I stretch the fibers with the active card from the stationary card and across its teeth. The carding action therefore happens in the space between the teeth of the cards. The quieter the strokes the better. Any force I need to use will put strain on both me and the fibers.

Figure 8.5. By carding gently, I catch the wool with the teeth of the active card so that the wool stretches between the teeth of the active and stationary cards in the space between them. With flat cards, I move the active card in straight movements. If the wool is tangled deep into the teeth, I have carded too harshly. I rest the stationary card in my lap and hold my stationary arm against my torso.

Figure 8.6. With curved cards, I move the active card in a rocking movement.

Figure 8.7. Transferring the wool: I hold the cards with both handles upward and move one card downward. This places all the wool on top of the other card. I do the same thing again with the wool ending up on top of the first card. This way I lift all the wool up from the bottom of the teeth and can make another round of carding. Note that I have folded the beard before I make the downward movement.

To keep a rhythm in my carding and consistency in my rolags, I decide on a number of strokes, usually between three and five. I stick to this number. If there are nepps or uneven sections in the wool, I troubleshoot at the earlier stages rather than card more (see below).

When I have done the strokes in my first pass, it is time to transfer the wool. There is most likely wool on both cards, and some slightly pushed down into the teeth. I transfer the wool, not so much for the transfer itself, but to lift the wool from the teeth and start fresh.

I hold the cards, handle sides up, in a *V* shape in front of me. I raise one of the cards and insert the top side about an inch (a couple of centimeters) below the start of the carding cloth of the handle side of the other card and gently stroke downward. This transfers the wool from the moving card to the still card (which card is active or stationary is irrelevant in the transfer). If the fibers are long and spill out over the top side of the card like a beard, I fold in the beard with the moving card against the still card as I start the transfer. I do the transfer one more time in the other direction—placing the top end of the card that now has all the wool against the handle side of the other card and make the transfer. The wool now lies airily on top of the first moving card. It is not necessary to make two transfers, but in my experience the wool becomes airier this way. I do another three carding passes with a double transfer between each one. The pattern may look like this:

- First to third pass: Four strokes (four rocking motions across the cloth with curved cards or four straight strokes across the whole cloth for flat cards) and a double transfer.
- Fourth pass: Four strokes and a double transfer, with the wool ending on top of the stationary card.

For each pass, the wool will become airier, more evenly distributed, and more structured. The fibers now lie like an airy, rectangular mattress—a batt—on top of the stationary card with the fibers somewhat aligned, structured vertically from top side to handle side.

Finish: Lift and Shape

For some spinning techniques (e.g., if I were to dress a distaff) and preferences, the batt is the end of the carding. I like to shape my carded preparations into rolags, though. I find rolags to be more stable and easier to spin an even yarn from than batts.

To shape the rolag, I place the stationary card (with the batt on top of the teeth) in my lap. I use my active card (teeth facing down) and my flat nondominant hand to shape the rolag into a Swiss roll shape:

1. I lift the wool from the top side of the stationary card from below with the top side of my active card.
2. I tuck the lifted wool with the edge of my flat hand and round it in, back of hand against the flat batt.
3. I lift the new, now rounded edge of the wool with the card and tuck the wool into the rolled shape with my flat hand and keep tucking and lifting all the way to the handle side of the stationary card.

Figure 8.8. By lifting the end and tucking it in repeatedly, I make a Swiss roll–shaped cylinder of my carded batt: a rolag.

Figure 8.9. I roll the rolag between the cards one more time to secure the end.

Figure 8.10. The finished rolag has the same width as the carding cloth, provided I have charged the wool within a frame on the carding cloth.

Figure 8.11. I hold the rolag into the light to check the quality.

4. To seal the rolag and give it a final shaping, I lift it between the handle side of the active card and the open palm of my hand, move it back to the top side of the stationary card, and roll the rolag gently along the stationary card once. Another way of sealing the rolag is to roll it back and forth between the wooden sides of the cards.

A carded rolag will be spun from its end, with the fibers coming in a spiral into the twist. It is also possible to roll the wool from the short end. This means that the rolag will be shorter and have the fibers arranged parallel to the length of the rolag. With this alternative preparation, I can spin the fibers end to end.

Now I do a survey of my rolag. I look at the length and the shape of it. It should be about the length of the card and hold together in the roll. I hold the rolag against the light and look for nepps, uneven parts, and tangles. If I find any, I go back a step or two and troubleshoot.

With the frame (amount of wool) and the counting (degree of carding), I will have rolags of even shape and fiber distribution. This paves the way to a smooth spin and an even yarn.

Troubleshooting Carding

Carding takes time to understand and learn. I have developed my way of carding for it to be gentle and enjoyable. Perhaps this way works for you, too. If not, try to find where the itch lies and adapt to a way that works for you. Here are some common challenges I have seen through my years of carding and teaching carding. Most of the solutions to carding challenges lie in the preparation, so a general advice is to go back to be able to move forward in your carding.

- **There are lumps and tangles in my carded rolags.** This is usually due to insufficient teasing. Go back a step and see if you can fine-tune your teasing.
- **The lumps and tangles in my rolags won't go away when I card more.** There is such a thing as over-carding. When the fibers are overworked, they will break and leave shorter fibers that bundle up into nepps, tangles, and lumps. This means that the more you card one portion of wool, the bigger the risk is of lumps. Keep to your chosen number of strokes and passes and refine your teasing instead (and have a look at the other points here in the troubleshooting list).
- **My rolags are layered and have bunched parts.** Go back to the dressing of the card. Is it possible that you have too much wool on the card? If you have overcharged the card, the teeth won't catch the whole mass of wool. This will leave layers and unevenly carded sections.
- **There are U-shaped bunches in my wool.** The "Us" happen if you card the wool over the outermost teeth of the handle side of the card. Try to stay within the frame when you dress the card. Also, try to start the carding movement below the handle side edge. When you transfer the wool, make sure to fold in the beard and start the downward movement below the edge of the still card.
- **I need to use force to card.** Have a look at your carding technique. Are you pressing the wool into the carding cloths? If so, it will be straining for you as well as the fibers. Try carding gently, brushing the wool between the cards rather than pushing it into the teeth. Tired arms can also be an issue of the cards. Some cards are tough and will cause straining. Try a different pair of cards to see if there is any difference.
- **The rolags are loose and won't close.** Shaping the rolag can be a challenge. The pressure from the active card in the shaping and final roll of the rolag needs to be firm enough to close the rolag, but light enough not to squash it. The shorter the fibers, the tighter the rolag needs to be to keep its shape.
- **My rolags are long and uneven (figure 17.3).** This might have something to do with the dressing of the card. As we card, the wool will fluff up and out as we bring air in between the fibers. Go back to the dressing and make sure you dress the card within a frame of about an inch (a couple of centimeters).
- **My arms get tired.** Check your position. Make sure to rest the stationary card in your lap and your nondominant arm against the side of your torso (figures 8.1, 8.5, and 8.6).
- **My legs cramp.** Check your position here as well. I see many of my students extending their toes to raise their legs. See if you can find a lower chair. If not, raise the floor by placing a book, a piece of firewood, or a rolled-up blanket under your feet. Another option is to lean the heels against the legs of the chair. I usually card beside my spinning wheel and place my foot on the treadle.

Combing

Why I Comb

Combing is about aligning the fibers so that they lie side by side in one direction. Combing wool in this way creates a worsted preparation. Aligned fibers prepare for a smooth and strong yarn. It is common to want all the tip ends in one direction and the cut ends in another. This way the scales of the fibers are facing the same direction and will reflect the light maximally. It also makes the yarn smoother. By combing the wool and aligning the fibers in the same direction in one long stretch into a top, I will get equal units of aligned fibers. While it is possible to spin a combed top in several ways, I usually aim for a worsted spun yarn (see chapter 10) when I comb.

What I Comb

I comb long fibers, preferably not under 4 inches (10 centimeters). With shorter fibers, the active comb can have difficulty catching the fibers from the stationary comb, and there will be more waste. Worsted spinning also generally requires length in the fibers. A combination of short and long fibers is also possible to comb.

Figure 8.12. The combing station is set up. In the image you see the staples, the finished top wound into a bird's nest, and the waste left in the comb.

Figure 8.13. For combing, I place the cut ends of the staples onto the tines with as little as possible showing on the handle side.

How I Comb

In chapter 7 where we discussed teasing the wool with combs, we established the terms *active comb* and *stationary comb*, along with *active ends* and *stationary ends* of the fibers. We will use these terms for combing, too.

The middle part of the combing process is the same as it is for teasing with combs, but the beginning and the end are different. You can use either mini combs or a combing station to comb (see also chapter 7).

Start: Charge and Direct

Just as when I tease with combs, I charge the stationary comb with picked locks. For combing, though, I arrange the staples in the same direction: I place the outermost parts of the cut ends onto the tines. I charge the staples to one-quarter of the height of the tines to avoid strain and tangles. The tip ends now point or hang from the front of the comb. As little as possible of the cut ends peek out on the handle side of the comb.

By arranging the fibers with all tip ends in one direction and all cut ends in the other, all the scales are facing one direction. This can make it easier for the fibers to align.

Middle: Align and Transfer

The next part of the process is the same as it is for teasing with combs—a perpendicular relationship

Figure 8.14. Just like for teasing or combing with mini combs, I start with a horizontal circle and keep it up until there is no more wool on the stationary comb.

Figure 8.15. The next circle is vertical.

between the combs, and combing in horizontal and vertical circles (see chapter 7). After a couple of rounds, I like to lift the wool slightly with my fingers or with the empty comb to distribute it evenly over the height of the tines, up to one-third of an inch (1 centimeter) or so from the tine tips.

When I start a new combing circle, I always catch the outermost active ends first, and then I move inward toward the stationary ends as the fibers detangle. This is not much different from combing long hair; you wouldn't start combing from the roots after a windy day. Instead, you would start at the tips and move toward the scalp.

When I circle my active comb, I make sure it glides parallel to the stationary comb. This way I catch as many fibers as possible. It is easy to come in at an angle, especially toward the end of the pass when the fibers left on the stationary comb are short, but stick to the parallel position.

The fibers I comb can be quite long, sometimes more than 8 inches (20 centimeters). I need to adapt my movements to the fiber length. The longer the fibers, the larger the movements I make. I need to make sure the fibers between the combs separate completely. So go for drama here—large circles for long fibers. For very long fibers, I may choose to use my combing station (preferably standing at an adjustable desk) rather than mini combs, since it makes it possible to move the combs in larger circles in a more sustainable way.

Finish: Lengthen and Nest

When the fibers are separated and I see no tangles or bunched up sections, it is time to remove the wool from the comb with the fibers aligned in a long and coherent top. This is sometimes called doffing. To do this as smoothly as possible, I want to start from the cut ends of the fibers. This way the scales will glide smoothly past each other instead of catching on to each other. We can compare this with combing and backcombing our hair. When we comb, we do it down toward the tips. This aligns the strands. If we

Figure 8.16. When the wool is evenly combed, I doff it from the stationary comb in one long stretch by gently pulling and wiggling.

Figure 8.17. I examine the quality of the combing against the light.

backcomb the hair toward the roots, the strands will instead tangle and build height. This is the scales working. So, if we started the combing process by placing the cut ends on the comb, we want to remove the fibers when the tip ends are on the comb. This means that we always remove the wool after an odd number of passes. For me that is usually after five or seven passes.

Before I doff the fibers, I need to decide which fibers I want where. I can choose to separate the fiber types (if I work with a dual coat) or keep them together. You can read more about this later in this chapter.

I grab the wool and pull—doff—gently, but only just under half of the length of the fibers. I take a new grip closer to the tines and doff the next section. If you are unsure of the fiber length your hands will feel it—whenever the pulling off goes easier, the fibers are telling you they are about to separate completely, so I would stop here and take a new grip.

I keep doffing until the remaining fibers are too short and nepps start showing in the fibers leaving the stationary comb. I have now removed a long top. It is probably wide in the starting end and thin in the finishing end. This is the perfect opportunity to survey my top. I hold it into the light and look for uneven sections, nepps, tangles, and vegetable matter. Depending on what I find, I may decide to change my method and perhaps add another couple of passes. You can read more about troubleshooting below.

To even the top out, I start at the wide starting end (the cut ends) and draft the top out evenly along

the length, while at the same time winding it loosely around the palm of my hand. As I do this, I keep looking for uneven and unwanted parts and remove these. When there are around 8 inches (20 centimeters) left, I catch the (tip) end between my index and middle finger and pull a loop through the wound part, so that the top is arranged like a bird's nest with an egg in it. If I follow this structure for all my tops, I will know that all cut ends are in the nest and all tip ends in the egg. When I spin the yarn, I can choose to start every top from the same end, resulting in all the scales in the yarn in the same direction.

Separating Fiber Types

When I doff combed wool with a dual coat, I can choose to separate fiber types and continue with different ways of preparing and spinning them. To do this, I doff only the outercoat fibers. These stretch past the undercoat fibers in the comb. If I hold the comb with the wool in the light or against a contrasting background, I can usually see the border where the undercoat fibers end and the outercoat fibers continue. When I grip the wool to doff, I need to place my grip outside of that border, to grip the outercoat fibers only. Once I have started doffing, the border will be easier to see.

When all the outercoat fibers are doffed, I even out the top and roll it into a bird's nest (see above). The undercoat fibers are now still in the comb. I pull these straight out, just as I do when I tease with

Figure 8.18. I even out the thickness of the top by drafting it slightly and winding it into a bird's nest.

Figure 8.19. If I want to separate undercoat from outercoat, I make sure I doff only the longest fibers (often outercoat). The shorter ones (often undercoat) stay in the combs and are now teased and ready for carding. The difference between the undercoat and outercoat fibers is clear.

Figure 8.20. Picked staples, teased and carded undercoat, and combed outercoat.

combs (see chapter 7). The undercoat fibers are now teased and ready for carding. If the doffed outercoat is just a very thin top, I may take a few tops and comb them together into one thicker group.

Combing Different Fiber Lengths and Planking

I can also comb a top with fibers with a variety of fiber lengths. When I doff the fibers, I need to make sure I grip both fiber lengths. This means I grip closer to the tines of the stationary comb than I would if I were to separate the fiber types. Still, no matter how thoroughly I grip, I will end up with longer fibers in the starting end of my top and shorter in the finishing end. If it is important to have a more even distribution, I can plank my top. This means that I comb it all again. I divide my top into a few shorter sections and add them back to the stationary comb. I prefer to do this on a combing station clamped to a tabletop. I recomb the newly arranged top two rounds and then doff again. Hopefully the fibers will be more evenly distributed after the planking.

If I am spinning a singles warp yarn, planking can be important even if I spin outercoat fibers only. If there is a variety in outercoat fiber length, the yarn will be weaker in the finishing ends of each top (where the shorter fibers are), and there will be a risk of warp threads breaking.

If the shortest fibers remaining in the stationary comb after doffing (or planking) are very short, I break my top before I get these in the top. I remove the short fibers just as I would for teasing with combs (see chapter 7) and card them.

What about the Waste?

No matter how thoroughly I comb, there will always be waste: short fibers, nepps, vegetable matter, and tangles. If there are fibers that are still usable and long enough, I gently pull them out and place them together with teased fibers for carding. I'm not a felter and have little use for the wool waste for felting projects (though this is a possible use for these fibers). The "waste" is never wasted, though. I use it in several other ways:

- I top my indoor plants with wool. This helps keep the moisture in the plants.
- I put it in my wellies (rubber boots) to felt my own wool inner soles.
- I place some wool waste inside my socks against the backs of my heels to prevent chafing. I always bring a small bag of wool waste for hiking trips.

Figure 8.21. If I have combed fibers of different lengths together, there is a risk that the longer fibers will stay in one end of the top and the shorter in the other. By planking the wool—combing it again in sections—I can even out the length distribution somewhat.

- I cover my outdoor plants and garden beds with wool. This keeps the moisture in the ground, prevents weeds from surfacing, adds to the soil structure with the help of the soil fauna who pull the fibers down, and helps fertilize the soil.

Troubleshooting Combing

Just like carding, combing takes time and practice to understand and learn. Do give it time. Still, even the experienced comber will face challenges. Here are some common ones with suggestions for solutions.

- **There are lumps in my wool.** This usually happens with overcharged combs. It is so easy to just add a couple more staples on the combs–they are so pretty! But too much wool on the combs will create a higher workload for you and less space for the fibers to separate. Try to stick with no more wool than to one-quarter of the height of the tines, or less for very dense staples. Sometimes the tip ends or cut ends can be tangled or felted, and it might be a good idea to tease them lightly with a flicker before combing. This will make the combing smoother and less straining and result in less lumps and less waste.
- **I need to use force when I comb.** Check where the wool is on the comb. Is it bunched up at the roots of the tines or spread evenly over the height of the tines? Distributing the fibers over the whole tine surface is often forgotten, which makes combing unnecessarily heavy.
- **It's straining to comb.** Check your charging. Have you placed the staples with as little as possible showing on the handle side? If not, recharge and see if the combing goes smoother.
- **There are tangles in my wool.** Sometimes when we comb, we loop the fibers on the stationary comb with the active comb. In the next circle, the active comb will catch loops instead of free fiber ends and tangle even more. This usually happens when we don't make the circle wide enough–the fibers between the combs don't separate completely. The longer the fibers, the wider we need to make the circle.
- **The top breaks when I doff.** Try doffing shorter portions, just under half the length of the fibers. Close your eyes and doff slowly. When you feel the fibers gliding more easily, they are about to let go. This is the time to take a new grip.
- **There is lots of waste in the combs.** There can be several reasons for this, from many of the points above. But it can also be due to the combing angle, especially in the horizontal circle. If you move your active comb at an angle in relation to the stationary comb instead of parallel, the comb will only catch a portion of the wool in the stationary comb (and risk catching and bending the

tines). Make the circle larger and make sure the active comb glides parallel to the stationary comb.

- **My nondominant arm gets tired.** Is your arm in the air when you comb with mini combs? That will put strain on it. Try resting your arm against the side of your torso.
- **I can't remember which end is where when I comb.** Talk to yourself, make a rhyme or song of your combing. Or write on your hands which end of the wool is facing which hand.

Can I Store My Prepared Wool?

In my experience, storing teased wool is rarely a problem, as long as it is done airily (but still away from pests). Prepared wool like rolags and tops is fresh produce, though. It will deteriorate with time. While it might be tempting to card or comb a large batch or a whole fleece before spinning, I generally only prepare the amount of wool I will spin that day or one bobbin or spindle worth of wool. What can happen if you leave your preparation is that time, gravity, people, pets, and wind will damage the preparation. While you may be able to protect the preparations from three of these environmental factors, time and gravity will meddle with the shape and structure of rolags and tops. The fibers will sink and push the air out, making especially rolags denser and more tangled. This process will happen faster with wool that is prone to felting. If you need to store your preparation, do it for a limited time and in a protected container such as a shoe box. Some wool will stay fresh for a day or two, some for a week. I invite you to experiment with this to find a way that works for you but remember that every fleece will be different when it comes to preparation freshness.

PLAY: SHAPE

Now is the time to play with cards and combs to find your way of preparing your wool. Your most important mental tools will be time and patience. Preparing wool with hand tools takes time to learn but is, in my experience, utterly rewarding. Therefore, be kind to yourself and practice a little every day. To be able to spin you will need to prepare your wool anyway. Try the wool preparation tools you have and see if you can borrow the tools you don't have. Play with preparing different kinds of wool, in different ways and with different tools, and be open to various results.

If things go wrong, and they will, analyze and troubleshoot to find what has gone wrong and where you may need to refine your technique (more of that in chapter 17). And remember, there is such a thing as low-quality tools, too. Take notes of what works well and what doesn't, which techniques you like and which you don't like. Perhaps you will find new ways or combinations that work better than the ones I have presented. This is your wool preparation, and you need to find a way that works for you.

When you practice, place your preparations in a row in front of you. The first rolag or bird's nest may look messy, perhaps the second one, too. But the fifth or the tenth will look better. As you practice, your hands will learn the movements and get to know the wool you work with. Look at your row of more than ten preparations and be proud! Look at the difference between the preparations in the rows. Try to find what you changed or refined in your method to improve the quality of the rolags and make notes of this.

Bonus invitation: Card wool that you haven't teased first and note the difference (see chapter 7).

Figure III.1. When I spin, my main focus is at the heart of the spinning. (Nettle shawl made by members of a Nepalese textile collective; see appendix 1.)

I Am a Spinner

Many times I have been asked about when I started spinning, and I tell the story of that box of wool that landed in my lap all those years ago. But when I once was asked when I became a spinner, I was given pause. I know I am a spinner, and I know how I started spinning, but what I hadn't thought of was when that shift from knowing how to spin to becoming a spinner happened, or how.

After that first spinning course, I kept spinning Pia-Lotta's fleece. I was a knitter and used both handspun and commercial yarns for my knitting projects. I learned about Kia, a professional wool classer with years of experience classifying wool in Norway. She was starting a fiber club with wool from rare and endangered Norwegian breeds, fleeces that she had hand-picked from the wool station. I enrolled, eagerly. In four parcels the wool came, samples of fleece with Kia's detailed notes attached to them, describing their characteristics, their background, and why the breeds and the fleeces were so special. I spun them all with love and gratitude to Kia and the generosity with which she shared her knowledge. When I had spun the last yarn from the last parcel, I wanted to do something special with the row of little test skeins I had spun with such reverence. I decided to knit a stranded colorwork vest where all the yarns got to shine, individually and as an ensemble. I included some of the yarn from my Pia-Lotta stash. After that I stopped buying commercial yarn. It wasn't a decision, just a natural transition from that point. With the help of Kia's deep knowledge of and love for wool I got the chance to dive into the characteristics of each fleece and make them shine. This was when I became a spinner.

Everything I create has its foundation in the wool and in the purpose of spinning. When I discover a fleece, I do so with the intention of finding its soul, of translating it into a yarn through my hands. When I make a textile, I continue that intention and make the yarn shine in the project.

Being a spinner is, to me, seeing the process from fleece to yarn as the main event, as the medium through which I experience all textile crafting and keep coming back home to. Spinning nourishes me. It is, along with writing, my main creative output, but it also gives me something more, a peace of mind, a moment to be in my spinning bubble and just breathe. In that flow of creativity, I find a joy that I don't want to be without.

Figure III.2. Yarn spun from Fjällnäs wool.

CHAPTER 9

Twist Model

I hold a perfectly rounded rolag in one hand, yarn in the other. Between them is a section stretching from free and airy to smooth and bound. My grip around the fibers is gentle—in the words of Judith MacKenzie, as if the rolag were a baby bird, tight enough to keep it in my hand, loose enough not to crush it. The yarn is in a soft hold in the other. As the twist moves from the twirling spindle into the yarn, I open up the twist gently, listening with my hands for the faint yet clear whisper of the wool: "Here I am. This is how long my fibers are; this is my crimp. This is the way I move." Yes, in that exact spot, in the heart of the spinning, where the fibers are allowed to move without disconnecting, that is where I hear the teachings of the wool.

The Twist Model is an illustration and a tool for spinning a more even yarn while being able to maintain ease in both body and fibers. In the Twist Model, we work in the point between spun and unspun and steer the twist toward the smoothest spin. In this chapter, I want you to learn how wool in general and your wool in particular behave in movement. I want you to learn to find that spot where the wool can convey information that will guide you to moving forward with ease and quality.

The Twist Model is essential in all the classes I teach. I use it to varying degrees with any spinning tool and any spinning technique. Take your time with this chapter. Be kind to yourself. It can be a pivotal point in your spinning. We will start with the very basics and take it step by step.

Basics

Take a piece of prepared fiber, a rolag or a piece of combed top. Hold it between your hands. Then, move your hands away from each other, still with the

Figure 9.1. A semi-stable section between unstable fibers and stable yarn.

Figure 9.2. When I move my hands in opposite directions, the unstable fibers in the rolag will come apart.

Figure 9.3. When I pull on fibers with twist, nothing will happen. The yarn is stable.

Figure 9.4. With enough twist for the fibers to move without coming apart, the fibers are semi-stable. This is where I will find the point of twist engagement and the heart of the spinning. If you focus on the dark fibers in figures 9.3 and 9.4, you can see the difference in twist angle—in figure 9.3 there is enough twist for the yarn to be stable, but in figure 9.4 I have turned the twisted part against the twist for the fibers to be able to move, but not back to twistless fiber.

preparation between them. Most likely this will result in the preparation coming apart. We can say that the fibers in a preparation are unstable.

If, instead, we put some twist in the middle of the preparation and move the hands away from each other, the most likely thing to happen is nothing. The twist prevents the fibers from coming apart. This is what we want in a finished yarn, right? It's generally not supposed to come apart with a gentle pull. In comparison to the prepared fiber, the spun yarn is stable.

We can illustrate this with a simple equation: fiber + twist = yarn.

Of course there is draft involved to get to the yarn, but we will come to that soon.

If, with the same piece of preparation, we untwist the twisted middle and pull, the fibers will come apart again. By reversing the equation, we get: yarn – twist = fiber.

Yarn and fiber, however, are not "on" or "off" features. Rather, they are the end points of a continuum with varying degrees of twist. Consider the range of twist between unspun Lopi-style yarn and a high-twist sock yarn. As spinners, we are in charge of this continuum. With our hands we can control the amount of twist we insert into or remove from the fiber, right between our hands.

Back to the preparation. It still has twist. If we turn the twisted part against the twist, not back to fiber, but to a place where the fibers can glide past each other without coming apart, we find ourselves with what we can call a semi-stable portion.

Some of us may compare it to driving a stick shift car—with a subtle balance between clutch and gas, we can control the car between movement and stillness. This is called dragläge in Swedish, and it's the word I use for the point where the fibers can glide past each other without coming apart. In English, I call it the point of twist engagement. This is not a point we can see or measure, though, and I will come back to this.

The Twist Model

There are equations to illustrate the basics of the relationship between twist and fiber. An equation, however, isn't sufficient to illustrate all the subtleties of spinning. Instead, I will use a simplified picture.

In the illustration, you see the unstable fiber to the left and the stable yarn to the right. The section in between is the semi-stable part, where the fibers can glide past each other without coming apart unless I let them. This is where we find the dragläge, the point of twist engagement. In the semi-stable part I can, just as with the clutch and the gas, steer the section between my hands between fiber and twist. At the point of twist engagement, the fibers can move—but not enough to come apart.

As I spin, I work with the semi-stable part all the time for some techniques and for troubleshooting and adjustment only in others (see chapter 11). With my fingers I can, using very subtle movements, come in and out of the point of twist engagement and spin smoothly.

Opening Up the Twist

Now, how do I get from stable spun yarn to the semi-stable part when I spin? Well, there are two ways. One way is to add length to the section between my hands. That means that the twist I had in the section to begin with will be distributed over a longer stretch of fiber. I work with this method in English long draw on a spinning wheel, a walking wheel, a floor spindle, a supported spindle, and sometimes also on a suspended spindle. Another way is to slightly untwist the yarn. I work with this method with any spinning tool and technique. I add enough length or remove enough twist to feel the fibers gliding in my draft, but not more.

To look at the untwisting method more closely, we can go back to the rolag between our hands, with the stable, twisted part in the middle. Let's say that I twisted the rolag clockwise. If I open up the twist, by rolling it counterclockwise, I will find the semi-stable part and the point of twist engagement. Let's play with this for a while. I hold the fiber that has been twisted clockwise and roll it against the twist between my thumb and index finger to come to the semi-stable part. To come back to the stable yarn

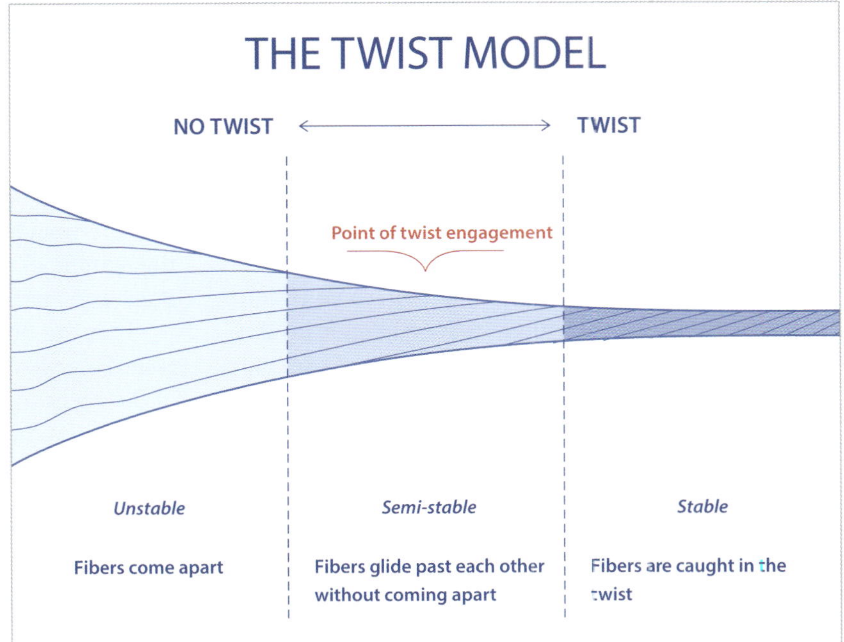

Figure 9.5. The unstable fibers (left) come into the semi-stable part (center) where the fibers can move without coming apart (unless I let them) before they become stable yarn (right). The point of twist engagement can be found somewhere in the semi-stable yarn. (Illustration credit: Isak Waltin)

Figure 9.6. The long draw in floor spindle spinning is longer than with most other spinning tools, and I need to be in control of the twist. I use my spindle hand thumb to subtly open up and let back the twist, coming in and out of the point of twist engagement. (Shawl pattern by Libby Jonson, knit in my handspun linen yarn.)

triangle, the semi-stable part, between my hands. The hand that holds the rolag is my fiber hand and the hand closest to the spinning tool is my spinning hand (learn more about hand roles in chapter 19). By rolling the spun yarn against the spinning direction with my spinning hand, I will open up the twist and allow the fibers to move without coming apart. This allows me to make a gentle draft.

Reversing Again

Now, the twist that I opened up is still around, just not between my hands at the moment. It bunched up inside my spinning hand when I opened up the twist between my hands. If I release the contact between my spinning hand fingers and the yarn, the twist will be let back into the yarn again. We have reversed the equation yet another time: fiber + twist = yarn. This time, though, I have drafted the fibers, so the section between my hands is longer—I have spun the yarn. I work like this, opening up the twist, drafting and adding the twist back into the yarn until I'm happy, and then rolling the yarn onto the shaft or bobbin. It's a very subtle movement, a roll, a draft, and a letting the twist back in. Meanwhile, twist is inserted

again, I let go of the pinch grip to add the twist back on. Back and forth.

The next step is to add draft. When I have rolled the yarn against the spinning direction to open up the twist, I can make the draft. I do not pull or use any kind of force. I just lean my hands in opposite directions to allow the fibers in the section between my hands to come apart slightly, which they will. With more opening up of the twist, though, I will have moved more toward the unstable part, and the fibers will separate fully.

Let's add a spinning tool to this and see what happens. I add twist to my yarn and have the spinning

into the yarn by either treadling the wheel or setting the spindle in motion. I can roll the twist open again. Back and forth.

Opening up the twist is easy to do with a slight roll of the yarn between the index finger and thumb of the spinning hand, but only up to a certain point. When there is too much twist in the yarn, a simple roll isn't enough. I may need to either roll the twist up several times or do the same action with the fiber hand, too. This is a sign that I need to work with slightly less twist when I spin. This doesn't mean I can't use the Twist Model for a high-twist yarn. I can spin with a lower twist, open it up until I'm happy with the evenness and thickness, and then add any extra twist I want.

Pinpointing

The Twist Model allows me to pinpoint problem areas, like an involuntary slub. I move my hands from their default spinning position and go closer to the area I want to unslub. With my hands in pinch grips on either side of the slub, I open up the twist in both ends to find the point of twist engagement. I make the draft and allow the twist back into the now evened out part of the yarn.

Benefits

What's in this for me as a spinner? Well, by unrolling the twist to get access to the draft instead of having to use more muscle power I save my muscles and can spin for a longer time without strain. With less strain on my muscles, there is a good chance that there is less strain on the fibers, too. If I have to pull on the fibers to find the draft, there is a risk I will break them.

By opening up the twist instead of pulling and pinching, I can also get a more even yarn. As I work with subtle movements to find the point of twist engagement, I can fine-tune the thickness of the yarn and focus on getting it evenly spun. This can lead to a smoother spin and an easier spinning rhythm. A light and even grip on the fiber also helps keep the thickness of the yarn even. An unevenly spun yarn due to pinching can lead to uneven twist and weak spots.

The process of finding the point of twist engagement can look different on different spinning tools and techniques, and we will look more closely at some of them.

Suspended Spindle

On a suspended spindle, the newly spun yarn holds the spindle. I hold the stable yarn in the spinning hand and the unstable fiber in the fiber hand. Between the hands is the semi-stable part. The distance between my hands usually depends on the length of the fibers. To access the point of twist engagement, I open up the twist—I roll the yarn against the twist between the thumb and index finger of my spinning hand.

The method works with the park and draft method (see chapter 14) as well as with the spindle spinning freely as I draft. With the spindle securely parked between my knees, I can work as slowly as I like, adding twist, parking, opening up the twist, drafting, and letting the twist back into the newly spun section. Add twist, park, draft, and repeat. Or, with the spindle spinning freely, set the spindle in motion, open up the twist, draft, and let the twist back into the newly spun section, flick or roll the spindle again and repeat, all the while the spindle spins freely.

Sometimes when I spin a woolen yarn (see chapter 10) on a suspended spindle, I add twist first and let it go into a good portion of the fiber preparation, a rolag in this case (figure 12.2). To access the point of twist engagement, I increase the length of the yarn—I move the spinning hand with the spindle in it an arm's length or less from the fiber hand. Since I have enough twist in the fiber, I know I will stay within the limits of the semi-stable part. I fine-tune by alternately opening up the twist and letting the opened twist back into the yarn as I spin.

Floor Spindle and Walking Wheel

With the floor spindle, such as the Navajo-style spindle, I work with both adding length and opening up the twist. I use a long draw of the same kind I would on a walking wheel (see chapter 10). This draft is often called English long draw. I roll the shaft against my thigh to add twist into the rolag. I make the draft by moving my fiber hand away from the spindle tip, spinning hand holding the shaft. I thereby lengthen the distance between my hands and allow the twist that was gathered in the rolag to travel along the drafted length. The whole arm's length between my hands is now semi-stable (figures 9.4 and 9.6). I add twist, open up the twist with a subtle roll with my spindle hand thumb, letting the twist back into the section, and work back and forth. When I have found the yarn thickness I want, I add the final twist and roll the spun section onto the shaft.

With the walking (great) wheel I do pretty much the same—I charge the rolag with twist by turning the wheel and then walk backward to add length to the yarn (figures 14.2 and 20.2). I can use my spinning hand to open up the twist somewhere along the length of the yarn, add more twist, and then roll the yarn onto the spindle by walking forward. With both spinning tools I can pinpoint a slub by placing my hands by each end of the slub to even it out.

Supported Spindle

When I spin on a supported spindle, opening up the twist is essential to keep the flow that for me is such a large part of the joy of supported-spindle spinning. I need to keep the twist low enough to be able to open it up with one subtle movement with my spinning hand. If that subtle roll isn't enough to open up the twist, I will need to stop the spindle and roll the twist open some more with my spindle hand. Opening the twist more or using more force to open it up will slow the speed of the freshly flicked spindle, giving my hands less time to spin.

So, with a supported spindle I flick the tip and move the spinning hand onto the yarn. I open up the twist between my spinning hand thumb and index finger, very subtly and multiple times, to use the spinning duration as efficiently as I can (figure 9.1). Too much twist will very quickly lock the yarn into a stable stage.

By opening up the twist, I allow a nearly constant flow of fibers into the semi-stable part. With the subtle and repeated opening up of the twist, I can keep the point of twist engagement alive as the spindle spins, before I roll the yarn onto the shaft and make the next flick.

Spinning Wheel

When I spin on a spinning wheel, I open up the twist with my spinning (front) hand. Since the treadling of the wheel generates a high speed, the movement needs to be even more subtle than with a spindle. If I find a slub, I generally stop treadling and unslub with my hands on both sides of the targeted area.

A Conversation

The point of twist engagement is not a point in the sense that I can measure it or mark it out visually. Rather, it is a circumstance where the characteristics of the fibers and how I adapt to them with my hands decide where that point happens. The circumstance can be about the fiber length and the way the fibers relate to each other in combination with how far apart my hands are and how much I open up the twist. Once my hands figure out the characteristics of the fibers, they will find the point of twist engagement. I will then understand how far apart I need to keep my hands and how much I need to open up the twist.

As I work with my hands in the semi-stable part, I keep a very light grip. My fiber hand holds the preparation gently. I usually keep a pinch grip on the yarn with my spinning hand, but always a light one. The hands are stable in their positions on the yarn and the preparation—the only thing gliding is the fibers along each other.

To find the point of twist engagement, I need to listen to the wool. It is the light grip that allows me to do that. As I hold yarn and fiber I get tactile feedback from them. I see this feedback as a signal that goes to my hands as well as between my hands. The signal carries the information I need of the fiber characteristics to find the point of twist engagement. As I hold the fiber and the yarn lightly, I can sense the signal and feel where the point is. Thereby I can understand how much I need to open up the twist and how far apart I need to keep my hands. In this way, the fibers communicate with my hands and the hands communicate with each other through the fibers. There is a pulse going between my hands and through the point of twist engagement, like the beating of a heart.

Going back to the illustration of the Twist Model (figure 9.5), it is only in the semi-stable part in the center that I get access to the signal from the fibers. The fibers in the unstable part are too loose, and the signal dies. The fibers in the stable yarn are caught in the twist and too tightly packed for the signal to come through. But in the semi-stable part, the signal has a chance to guide my hands to the point of twist engagement, and only if I keep that light grip—as soon as I pinch my fingers or pull the yarn, the signal gets cut off.

CREATE: CLOSE YOUR EYES

Now it's your turn to trust your hands to listen to the wool. Close your eyes and open up the twist with a light grip on whatever you are spinning, or just on a rolag between your hands. Allow your hands to really take in the signal from the fibers to find that spot where the fibers glide past each other without coming apart, the point of twist engagement that I see as the heart of the spinning. Make the draft and allow the just opened twist back into the newly spun section. Keep going, explore, and see what happens. Make some notes if you wish.

CHAPTER 10

Spin, Ply, and Finish

In this chapter, I will present my thoughts about woolen and worsted spinning, what they are, and why we spin with a woolen or worsted technique. We will also look at plying and finishing the yarn.

Woolen and worsted yarns are two end points of a spinning spectrum with a large section of semi-woolens and semi-worsteds in between: from warm to strong, from lofty to defined, from light to drapey. The main difference between woolen and worsted spinning is the relationship between the draft and the twist. The woolen and worsted yarns will be either soft and airy or strong and dense (figures 6.3 and 12.3). The spinning spectrum in between includes variations of these end points.

There are many opinions about what true woolen or true worsted yarn is, and to me that's just what they are—opinions. What someone else thinks is a rule is not important to me. I want to understand why I spin with a certain technique and explore my hows from there.

Woolen Spinning

The aim of a woolen yarn is loft, warmth, and lightness. It's not a particularly strong yarn, and it doesn't have the most shine or drape. Spinning a woolen yarn is quite fast and requires less wool than a worsted yarn. Because of its construction of short fibers in a lofty preparation, a woolen yarn can pill.

Woolen spinning is about drafting against the twist, letting the twist into the fibers before or during the draw. This means that when we make the draw (in woolen spinning often called a long draw), the

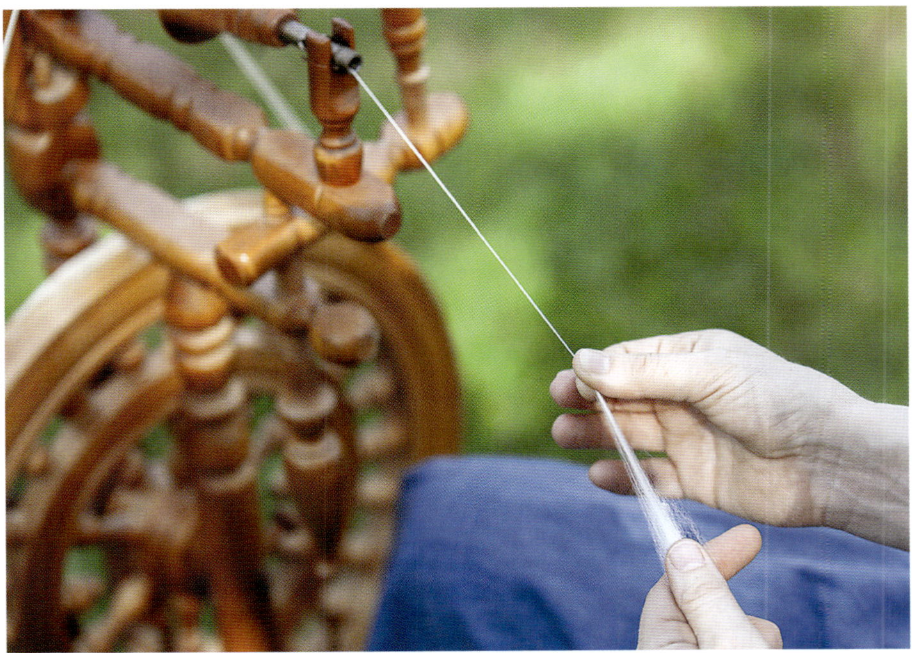

Figure 10.1. I add the twist after I have made the draft in worsted spinning. The spinning (front) hand controls the twist and lets it into the fibers after I have made the draft.

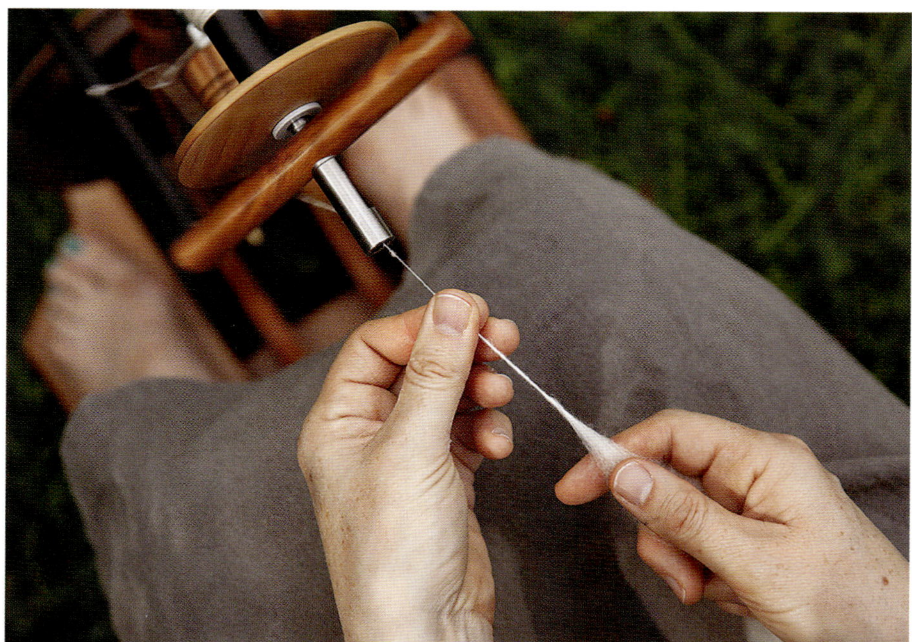

Figure 10.2. I let the twist into the fibers before or during drafting in woolen spinning. With the spinning (front) hand in a light grip, I can control the twist coming into the fibers so it doesn't lock them.

twist is in the fiber. The undrafted fibers that get jumbled by the twist make room for air. The resulting yarn is lofty and generally soft, and the air that has been trapped between the fibers makes the yarn warm.

A woolen spun yarn can mean many different qualities in the woolen end of the woolen-to-worsted spectrum. Several factors are important: the fibers, the preparation, and the drafting technique. A woolen spun yarn requires a woolen preparation, and I use hand cards for this (see chapter 8). The easiest fibers to card and spin woolen are fine and short, perhaps between 2 and 4 inches (5 and 10 centimeters). They may also have a high crimp. As I card, I arrange the fibers airily and somewhat aligned, from handle to top side of the card. The rolling of the rolag arranges the fibers in a cylindrical shape. When I spin the rolag from its short end, the twist enters the rolag in that circular arrangement, and the fibers place themselves jumbled, but still airily and evenly, in the yarn.

There are many techniques to spin woolen yarns, and I will share my favorite, which is often called an English long draw, a technique I have touched upon in chapter 9 from a Twist Model perspective. On a floor spindle, such as the Navajo-style spindle, I like to charge the whole hand-carded rolag with twist before I make the draw. I hold the rolag very loosely in sort of an O-shaped fist. I roll the shaft several times up my outer thigh to insert twist. When I feel the rolag squirming in my hand, I know it has enough twist. I make an arm's-length draft from the whole length of the rolag. I now have a lightly twisted pre-drafted ex-rolag between the shaft and my fiber hand. The thumb of my spindle hand is lightly resting on the yarn. Now I can work in arm's-length sections of the pre-draft: I wind some of the pre-draft onto my fiber hand for storage and open up the twist (see chapter 9) of the section that remains between my hands to fine-tune it. When I'm happy with the evenness and thickness of the section, I add the final twist and roll that spun section onto the shaft. I unwind another arm's length of pre-draft from my hand and fine-tune that one until the whole length of the pre-drafted rolag is spun and wound onto the shaft. The long draw feels pleasant, and I enjoy watching the rolag turn into a pre-draft right before my eyes (figure 9.6). This technique can also be adapted to other spindle types, such as supported and in-hand spindles, but also suspended spindles (see chapter 12 and figure 12.2).

Spinning English long draw on a spinning wheel has a smooth rhythm. With a pinch grip on the yarn, I charge the yarn with twist for a set number of treadles. Then I let go of the pinch grip, make an arm's-length draw with my fiber hand, and add twist for a second set of treadles. On a walking wheel I walk in a certain pattern—I set the wheel in motion to charge twist, take three steps back for the draft, add twist, take one step to the right to change the angle of the yarn, and take two steps forward to roll the yarn onto the spindle (figure 20.2).

Spinning a woolen yarn requires trust—in the twist to hold the yarn together and in yourself to feed it with fiber. The twist locks the fibers in the draft as long as the draft isn't made faster than the twist runs. Spinning a woolen yarn also relies on the spinner to feed equal amounts of fiber into the twist. This can be tricky in the beginning, but with practice, the hands will learn the rhythm of feeding, adding twist, and drafting. Trust this intelligence in your hands (see chapters 18 and 19). Furthermore, woolen spinning requires trust in the yarn to become consistent through your hands. The technique with air trapped between jumbled fibers can result in uneven spots. With practice and thorough wool preparation (see chapters 7 and 8), these will be fewer and more even, but they will also even out in the ply and in a textile.

Worsted Spinning

The aim of a worsted yarn is strength, shine, and drape. A worsted yarn will not keep you as warm as a woolen yarn and is not as light and soft. Spinning a worsted yarn takes time. In worsted spinning, we make the draft first to align the fibers tightly, and then add the twist. Air is pushed out, and the resulting yarn has the fibers quite densely packed and aligned (figure 10.1). A worsted spun yarn is a good candidate for warps. In knitting, a worsted spun yarn will have a high stitch definition.

A worsted yarn can represent many qualities in the worsted end of the yarn spectrum, just as the woolen yarn in the other end. Fibers, preparation, and drafting technique are keys to the yarn quality we are looking for.

To spin a worsted yarn, we need a worsted preparation—just as we align the fibers when we spin, we align the fibers in the preparation. I do this with combs, and preferably with longer fibers, at least 4 inches (10 centimeters) (see chapter 8). The less crimp the fibers have, the higher the chance of a dense and shiny yarn.

The combed preparation aligns the fibers, and with the drafting of the fibers from the combed top, they are arranged end to end in the yarn. If I have kept track of the tip and cut ends of the staples as I have dressed my combs and doffed the wool off, all the cut ends are facing one end and all the tip ends the other. This means that the scales of the fibers also face one end, giving maximum shine in the yarn. The direction of the fiber ends feeds a lively discussion about dos and don'ts: I like to spin from the cut ends. I think the fibers draft easier that way, and they feel smoother against my fingers when I spin (see the back-combing analogy in chapter 8). Some say this makes the fingers smooth the yarn with the scales rather than go against them, aiding in keeping the yarn sleek. Others say this method will ruffle the scales in the plying and make the yarn fuzzy and, therefore, favor spinning from the tip ends instead. Experiment and see what you prefer.

To keep the fibers aligned and dense in the short draft, I need to prevent the twist in the spun yarn from entering the draft, using a light pinch. When the drafted fibers are straight between my hands, I let go of that pinch and allow the twist into the drafted fibers, moving the spinning hand to meet the fiber hand while smoothing the spun section down. I move the fiber hand up the preparation to draft a new section of fibers. When the fibers in the drafted section are straight, I ease the hold in the spinning hand again to let twist into the new section. I let as little air as possible in between the fibers and keep them straight before I allow the twist in. This will give me a dense and strong yarn with lots of shine.

I like to spin worsted yarns on my spinning wheel and on suspended spindles. The weight of the spindle adds to the pull of the yarn, thus straightening it and perhaps squeezing out some extra air (figure 12.1).

The Semis (in Between)

There are several ways to create yarn in the spectrum between woolen and worsted. These yarns are generally called semi-woolen and semi-worsted. Exploring characteristics typical for woolen yarns in a worsted preparation will move the yarn from the worsted end of the spectrum toward the woolen, and vice versa.

A semi-woolen yarn can be carded rolags spun with no twist between the hands, with a mix of shorter and longer fibers or with a low-crimp wool. The yarn will have less air and warmth than the yarn spun on the woolen end of the spectrum but will be stronger and shinier. A semi-worsted yarn can be a combed top spun with twist between the

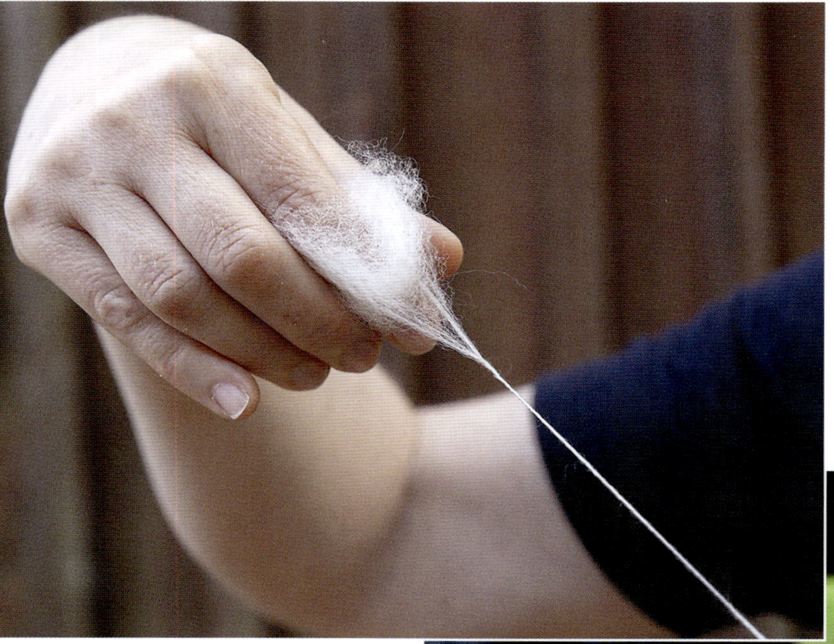

Figure 10.3. Spinning from the fold is a way to hold the fibers. In this example, I allow twist into the fibers before and during the draft.

Figure 10.4. Here I draft the fibers before I allow twist in. I can't keep the draft completely out of the fibers since I don't have that level of control in this kind of grip, though.

hands, with a mix of fiber lengths or with high-crimp wool. The yarn will not be as strong as the yarn on the worsted end of the spectrum but, instead, softer and warmer. The semi-woolen and semi-worsted yarns can be used with a wide variety of textile techniques for a wide variety of textiles. The opportunities to play are many and inviting.

Some wools are suitable for spinning over the fold, preferably long and fine ones. Spinning over the fold means that the spinner folds a section of fibers over the finger of the fiber hand and drafts from that fold, from the center of the fibers. The fibers enter the yarn from a wide angle and strive to unfold, resulting in an airy yarn. Spinning over the fold isn't a spinning technique in itself. Rather, it is a way to hold the fibers. They can be spun with either a long or a short draw.

Revisiting the Twist Model

In chapter 9, we looked at the Twist Model, a tool to open up the twist for smoother spinning. I open up the twist throughout woolen spinning: every time I make a draft, I subtly roll the yarn against the twist and add twist back into the yarn.

As the twist isn't in the fiber for worsted spinning, there is no twist to open until the yarn is already spun. However, I use the Twist Model when I want to smooth out a slub or tangle (for worsted and any kind of spinning). This is where I move my hands to frame the section I want to improve and roll the yarn against the twist with both hands. When I find the point of twist engagement, I can lean my hands away from each other and let the fibers glide until the section has the thickness and orderliness I want.

Plying and Finishing

A singles yarn works perfectly as a yarn, but it does have energy: the yarn you have twisted in one direction naturally wants to untwist. If you hold up a singles skein, the strands tend to twist back on themselves and create pigtails. That can be pretty and fully functional in some contexts but challenging in others.

In knitting, a singles yarn can make the fabric biased—again, since the energized yarn wants to untwist. This can be remedied by knitting in a balanced structure such as ribbing, garter, and moss stitches. In chapter 18, you can read about a pattern I have published with singles yarns with a balanced knitting technique. A singles yarn usually works well as a weft yarn in weaving. It works as a warp yarn, but in my beginner experience the energy in the threads may lead to tangling and breaking, especially if they are unevenly spun. A singles yarn is not as strong as a plied yarn and is not ideal for socks, elbows, or cuffs that wear a lot.

If you want balance and strength in your yarn, plying is a good option. I ply most of my yarns that I don't use as weft yarn. The twist in a plied yarn needs to be in the opposite direction compared to the two or more singles in it, unless you are going for designer yarns. The twist in a plied yarn also needs to have the same amount of twist as the singles for it to be balanced. One way to find the proper plying twist is to measure the twist angle in the singles and use the opposite angle for the ply. That has never really worked for me, so I have other ways to find my optimal plying twist.

When I ply, I do look at the twist angle, but in the plied yarn and without a measuring tool. If the singles in my yarn are spun clockwise, the ply needs to be counterclockwise with the same amount of twist, the mirror angle of the singles. This means that, in a balanced ply, all the individual fibers in the singles are parallel to the direction of the yarn. This way I don't need to measure angles. I just look at the ply, perhaps through a magnifier, and check that the fibers are aligned with the direction of the yarn. The strands in the underplied or overplied yarn have an angle.

An additional way to determine balance in a ply is to look at the shape of the yarn. A balanced plied yarn usually has visible bumps where the singles are twisted around each other (see above and figure III.2). In overplied yarn, the bumps are flattened and not visible at all. The singles in the underplied yarn are limp and look more like individual strands loosely wrapped around each other.

Figure 10.5. The middle strand shows a yarn in balance. The individual fibers are parallel to the yarn, and the bumps are defined. The left yarn is underplied, and the strands look loosely wrapped around each other. The fibers are tilted. In the overplied sample to the right, the fibers are tilted in the other direction, and the bumps are flattened.

I use different methods to arrange the singles for plying. For plying on a spinning wheel, I use my (Lazy) Kate that can manage three bobbins. If I ply on spindles, I may have stored my singles on empty toilet rolls that I use in the Kate. It is also easy to make a Kate from a shoe box or a paper bag: Pierce holes in the long sides of the shoe box (or the paper bag with the walls folded in half into the middle, handles cut off). Slide long knitting needles or barbecue sticks into the holes and thread the toilet paper rolls onto them and ply from there. If the singles tangle, place a plastic bag underneath the rolls as a tensioner. If I have spun my singles on separate spindles, I can slide them into the holes and ply straight off the spindles. This usually works best with supported spindles where there is no hook in the way. Another way is to wind the singles onto pebbles, tennis balls, or marbles and let them roll on the floor or in bowls. The weight keeps the singles tensioned.

When I have made sure the individual fibers are aligned with the yarn, I keep plying, checking the shape of the yarn regularly. I keep the singles separated between my fiber (plying) hand fingers and evenly tensioned.

After plying, I wind the yarn onto a niddy-noddy and tie off. I finish the skeined yarns by soaking them in warm water with organic shampoo. The soaking evens out the twist and allows the yarn to settle into its final shape. After rinsing in two or three waters of the same temperature (and perhaps white vinegar in the last rinsing water to neutralize any alkalinity), I roll the skeins into a towel and step on it to squeeze out as much water as possible. I place the skeins over my arms and pull outward to set the twist. Another way to do this is to whack the skeins on the floor or a door.

CREATE: DIVE

Now, try different drafting techniques, from the end points of woolen and worsted, and the semis in between. Spin the fibers and preparations that are typical for a spinning technique, but make sure to do the opposite, too, and note the difference. Explore tools, techniques, and wool types and dive in! Keep an open mind and see what you get. Make plying samples until you like the look of the yarn, ply it, and finish.

CHAPTER 11

Be Kind

Learning a new technique or tool, or diving deeper into the ones you already practice, can be overwhelming at times. Let's take a break and be kind to ourselves in our endeavors to create the most beautiful yarns. In this chapter, I want to encourage you to look into your wool to find the answers to your questions. I want you to listen to your body and trust your own capacity to learn and create, ground in what you know, and explore from there.

You Are the Expert

When I teach, I have both experienced and beginner students in my classroom. All levels of students get frustrated at one time or another. My reply to these frustrations is to practice patiently and curiously, and to be kind. There is no shortcut to spinning and preparing wool—you need to practice to understand the technique. In chapter 8's exercise "Play: Shape," I encouraged you to practice carding and combing, placing the rolags or birds' nests in a row. This is a gentle way of reminding yourself that you are, in fact, making progress. It might not show from the first to the second preparation, but probably from the first to the fifth. And between the first and the fifth, your hands and brain have practiced this new movement pattern many times and started to recognize and understand it better (see chapter 18). They might just not have shared the news with you yet.

There are many established "truths" about how to prepare and spin a specific breed and how to—or how not to—execute a specific technique. These truths may be true for some spinners in some circumstances, but other techniques can be just as true for you in yours. And for other spinners, they might just not work. You have a unique skill level, a unique set of tools, and a unique context. Only you know what works for you in each situation. By having the wool

Figure 11.1. I keep a gentle grip on yarn and fiber to avoid uneven yarn and strained body.

going through your hands over and over through the preparation, you know it well. In fact, with the tools and experience you have, you are the expert on this fleece and on the yarn you are creating.

Go Easy to Go Far

To be able to spin for a longer time without strain, we need to look at our tools, how we sit and place our bodies when we spin, and how much time we spend in different parts of the process. I spin for joy, and I want to keep the spinning joyful. Perhaps you do, too. Therefore, we need to take responsibility for our spinning comfort. Listen to your body and find a path for your spinning process that works for you.

The Grip

One thing I notice when I teach is the hold-on-for-dear-life grip on tools and wool. I see many sweaty and felted lumps of wool in inexperienced fiber hands and ask the students to look at how they hold the fiber. I do it myself when I learn something new or come back to a technique I haven't used for some time. After a while my grip loosens, and I find a comfortable and confident way to grip. My theory is that when something is new to me and I can't get a grip on the technique cognitively, I compensate by gripping the tools and the material physically. Once we trust our hands to know what they are doing (or trust that they will eventually) the grip lightens (figure 11.1). But before that, I as a teacher need to remind my students to lighten their grip and inspire them to take breaks and shift techniques.

Levers

A long lever will take more muscle power to lift and will put more strain on the joints. When we spin, there are ways to keep the levers as short as possible, without interfering too much with the motions and the technique. A tool held in the passive hand can be supported in the lap, and the arm can rest against the torso (figures 7.5–7.8), rather than holding the tool in the air and the arm like a wing. An active hand can work in smaller movements with wool that isn't too long. That way the motions don't force the joints too much beyond their neutral angles.

We have all been there, though—lifting the fiber hand in suspended spindle spinning higher and higher, just one more section, until we are ready to climb up the stairs because we, for unknown reasons, want to spin as much as possible before we roll the yarn onto the shaft. Or we hold the fiber hand in supported-spindle spinning way above the shoulder before we roll the yarn in. I know I have, just as I have pulled my arm way back for an English long draw on my spinning wheel. None of these techniques are necessarily bad for a limited time, but it is when we keep repeating the extreme executions that we put our bodies through unnecessary strain. So do take breaks, shorten the levers where it is appropriate, and perhaps sit on a swivel chair for your longest long draws.

Now I See You; Now I Don't

Have you looked at the lighting at your spinning space lately? Would your eyes be less strained if you directed a spotlight at your hands or your tools? Or perhaps used a flexible neck light or magnifying glass for specifically demanding techniques? I like to spin in the living room in the morning, with natural light coming from behind me and from the left. But in the evening or in the winter, I use a double spotlight close to the wheel and three background lights from different angles across the room. We don't always notice the lack of light until we turn on that extra light; then we experience a huge difference. The neck releases and the brow relaxes, just like the shoulders unexpectedly sink when the oven fan silences and you realize it had been on.

Sometimes I have sensed that something was insufficient, bothersome, or causing strain, but it was neither the lighting nor the glasses. After a while, I realized it was the towel in my lap that lacked contrast to the yarn. A natural linen towel gives me too small a contrast for a natural white yarn. Now I make sure to use a darker background for a light yarn and a lighter one for a dark yarn.

Tools

When I took my first spinning lesson (see "A Box in My Lap" on page 1), I bought the only suspended spindle that was available, a 3-ounce (90-gram) piece with a 12-millimeter-thick shaft and a fat hook. It functioned as a spinning tool, but it was neither comfortable nor joyful. I have no time for tools that impede my spinning process. Many of my students share this story of the heavy "beginner's" spindle. It may of course be a replica of a traditional spindle or an archaeological find. Personally, I believe many of these were made by someone who isn't a spinner

Figure 11.2. Contrast between yarn and background is vital for a comfortable spin.

themselves. I also believe that this type of spindle is one reason why many beginners abandon spindle spinning: it's just not joyful, and they haven't experienced spindles that are.

When I started to teach supported-spindle spinning, I talked to a Swedish professional wood turner and asked him to make supported spindles for my classes. He said he would, but not without knowing how they were used. Turning a beautiful object is one thing, but turning a tool is a whole other story. A tool is used for something, and in that it needs to be created for that specific use and for the comfort of the user. The aesthetics are just a bonus. I showed him, and he made some prototypes. My students and I later explored grip, friction, height, balance, and shape, and gave the maker feedback for his next batch. Today that wood turner, Björn Peck, makes all the spindles for my supported spindle courses. They spin like rockets. His spindles are in high demand, and he sells them to spinners worldwide.

When I look at tools, I always look for specific details, and I sometimes have a dialogue with the maker. The function of the tools we use needs to be an ongoing discussion. If something is a nagging issue, we need to find the problem and prevent it to the best of our abilities. We can't always buy new tools, but we can be aware of any features that can be causes of that problem and make sure we don't strain ourselves. For more ways to avoid strain, look at the troubleshooting sections in chapter 8. You can read more about the tools I use and some features I look for in appendix 2.

Parallel Projects

Coming back to woolen and worsted spinning, there are important considerations. Spinning woolen is faster than the worsted technique. Repeated arm's-length long draws, however, can be a strain for the shoulder. Pinching the yarn in a worsted technique can be straining for the hands. When I separate a fleece and prepare and spin undercoat and outercoat fibers differently, I love to alternate between the yarns. I card, spin, and ply one skein of woolen yarn and then comb, spin, and ply one skein of worsted, and keep alternating through the entire fleece. That way I get both mental and physical variation.

Other things to alternate include

- picking, teasing, and carding;
- spinning high- and low-crimp wool;
- spinning on spindles and wheels;
- spinning with left and right hand as spinning hand (see chapter 19);
- spinning position; and
- the place I spin.

Pause and Move

I usually stand or walk when I spin on a suspended or in-hand spindle. It gives my body freedom and the spinning process a rhythm. It may sound scary to spin while walking, but with a little practice, it is a wonderful experience. Just start slowly and find your pace (figures 14.1 and 15.1).

I generally stand when I use my combing station at my adjustable desk. My body has more freedom, and I can stand closer to the tools and still reach longer. Sitting down at the combing station makes me feel confined and inadequate in my movements. When I use mini combs, I love sitting, perhaps since there is nothing in front of me to obstruct my movements.

Even if you do have a combing station or an adjustable table, do take breaks. Dance around; place your tools far away from where you sit so you need to get up to get them. Go get a drink of water and just look to the horizon for the giraffes. Our eyes were not created to look down into our laps all the time. Instead, they are adapted to look for danger on the savannah. So, I stand up and look out the window for the giraffes. Not that I think they are any danger to me in Stockholm, but I prefer them to lions.

During the pandemic, I worked from my home office for two years. Every day I did a short movement series on Microsoft Teams with my colleagues. They quickly named it the "Sledge and the Scoop." In the series, we go through the four movements of the spine: rotation, flexion, extension, and lateral flexion. This is an excellent movement series for spinners, too, especially if we have been sitting for a while making repeated motions.

START WITH THE SLEDGE (ROTATION)

As shown in figure 11.3, stand up with your feet hip width apart. Start to rotate your upper body left and right with your hips facing forward, so that the rotation happens in the torso (rather than in the legs). Let your arms swivel along, perhaps knocking into your body before you turn back. After a while, let the movements get smaller and finally become still.

Figure 11.3.

NOW, FOR THE SCOOP

1. Stand with your feet shoulder width apart. Place your open palms, pinkies touching, in front of your body. Your hands now form the scoop (figure 11.4).
2. Bend your legs as much as is comfortable, pointing your scoop down (flexion), turn back up at your lowest point, and scoop up to standing (figure 11.5).
3. Turn your scoop toward you and make a loop in front of your body. Finish the loop and stretch your arms overhead (figures 11.6 and 11.7).
4. Lean your long body to one side, through center and then to the other (lateral flexion) (figures 11.8 and 11.9).
5. Come back to center, bend the elbows into a goal post and puff up your chest (extension) (figure 11.10).
6. Move your bent arms forward and your hands back to the scoop, pinkies touching again.
7. Repeat the scoop as many times you see fit.

Figure 11.4.

Figure 11.5.

Figure 11.6.

Figure 11.7.

Figure 11.8.

Figure 11.9.

Figure 11.10.

Learning Modes

The classroom is set up for a new course. In just a few moments they will come in, brand-new students who trust me to guide them through their spinning journey. They will all have different learning styles that I need to find and decipher. Some will learn quickly, some slowly. Some will be eager and stumble, others will need to talk to themselves for the extra feedback loop. A few will have low confidence in their ability to learn. Through my years of teaching spinning, I have made myself a well-structured toolbox of techniques to get through to spinners with different learning styles and unique contexts, but there are new challenges in every classroom. The student who kept struggling and I kept focusing on until I realized she needed to mutter and get frustrated on her own for a while until she was ready for my guidance. Or the student who wrote down every word I said. She was very aware of what methods worked for her. I just needed to realize it, too. In chapter 16, I give more examples of situations where students have found their own solutions that suited their individual needs.

I want each student to walk out of the classroom when the course is over with something more than just spinning skills: a sense of achievement, of joy of having learned something new. For that to happen, I need to understand their learning styles and find a channel and connection to each of them.

As a teacher, I find it important to get to know the foundations of my students and their respective capacity to explore. I want them to find that dynamic between the points of grounding and exploring that makes them smile and sing "Aaahh!" as they see their progress. I want to be there, right beside them as they learn. Look, they're coming! I wonder what they will teach me this time.

CREATE: FLOW

At the end of one of my recurring courses, "A Spindle a Day," I encourage my students to let go of any demands or expectations of the results and just spin for the process. In this exercise, I would like you to spin for you. Choose your favorite wool preparation or spinning technique and your favorite tools, and place yourself comfortably. Spin (or prepare your wool) for the sake of the process, not for the result. Find a rhythm that works for you. Don't bother about slubs, tangles, technique, or rules. Just enjoy your moment.

CHAPTER 12

Break the Rules

Go Wild to Expand

Look at me! Aren't I the most beautiful fleece you have ever seen? Mmm, my curls, my shine, my sweet and floofy dandelion puff undercoat, outercoat shinier than the finest silk. Come dance with me; spin me around and rejoice in all I have to offer. I bet you have never spun anything like me before. How would you make me pretty? Oh, don't say you will do the usual stuff–comb my outercoat and card my undercoat–do something new, something wild! See me as the precious pearl I am and marvel at my reflection. Perhaps you will find magnificence. Trust me, it'll be worth it. I can make magic with my fibers, you see; I have what it takes to become a stellar yarn. Comb my short fibers, card my long, and blend me with specks of joy and freckles on a summer morning. I want you to find sparkle, sass, and sizzle. I want you to giggle from your toes at the beauty of creating this yarn. Take all the rules you ever learned, toss them into the air and watch them crumble and show their true nature as they fall. Do it to learn, to get to know my limits as well as your own, to find my range of possibilities and yours. Perhaps you understand deep in practice what you knew vaguely in theory. Save the new insights like sweets in a crystal bowl and be reminded of them every time you pick one up to watch it shine. This is your chance to put all the rules to the test, turn them inside out, look at them from underneath, and widen your horizon to levels you didn't know you already love. ✻

You may have heard it before: You need to know the rules before you break them. For me, that isn't necessarily the whole truth. I need to feel the implications of the rules, sometimes painfully, to understand them, and from there decide whether I want to break them. With a physical memory and the understanding of what happens when I break a rule, I can experiment with where I break it or bend it and decide where I am open to cutting corners. In this chapter, I want you to go wild and experiment in an endeavor to understand the wool and to trust your experience.

Making Physical Sense of the Rules

It's so easy to just not break the rules, to not question them. You have been told to do something in a particular way. It has become a truth, and you have never challenged it. Sometimes, though, we don't know why it is a rule. Perhaps it originates in a specific geographical or cultural context. For me, rules need to make sense, in my frame of reference, and usually physically. I need to feel in my process how a rule is smooth or itchy or just has a bad taste of coriander. Something can make sense to me when I hear about it, but it's not until I have tried–and often failed–that I can understand why and make decisions from there.

In 2021, I spun a yarn with nalbinding in mind. I used vadmal-type staples (see chapter 2) of Åsen wool. But just as much as I had a textile technique and project for the yarn in mind, I wanted to spin for the process. I picked a newly acquired suspended spindle and started spinning my carded rolags. To give the yarn extra loft, I allowed the spinning spindle to dangle from the rolag as the twist traveled up the rolag and contorted its cylindrical length, before I made the draft. It was a technique I hadn't used before, and I got totally smitten by it. Making the draft with the twist deeply rooted in the rolag was a joy, and I found myself giggling as I set the final twist to the yarn. I shared a short clip of the technique on social media. Most of the comments in the thread showed the same infatuation as I had experienced. Among the comments, one puzzled me, though. Someone commented, "Allowing the twist into the supply?! Blasphemy!" I didn't know there was a rule. To me, it was just a yummy way of creating a yarn. After I had read the comment (several times), I enjoyed the technique even more.

As spinners, we can explore a rule we learn about, not much different from deriving the Pythagorean

Figure 12.1. Some people find spinning in the grease ghastly; others, like me, find it joyful. Here I spin straight from the newly shorn fleece of Parisa the Dalapäls sheep, who is standing on the shearing table by my fiber hand.

theorem: we accept it when we use it but understand the implications of it only when we challenge it. In chapter 8, I talk about fiber length for carding and combing: I don't comb wool with a fiber length shorter than 4 inches (10 centimeters), and I don't card wool that is longer. But is such a rule or recommendation always true? What happens when I comb 2-inch (8-centimeter) fibers? Are they difficult to catch, or won't they stick to the tines at all? Can I bend the rule if I use smaller combs? And what happens if I card 6-inch (15-centimeter) fibers? I know they can double up in the rolag, but what if I keep them as a batt? At what conditions do I need to switch from a wider spaced set of combs or cards to a smaller one? I need to try what may instinctively feel odd to find out what feels right.

Trusting My Experience

The context you work in is your guide. To learn what works and what doesn't, you may need to challenge

Figure 12.2. The blasphemy of adding twist to the fiber supply is one of my favorite methods of spinning with a suspended spindle.

the rules in the context of your skill level, your tools, and the fleece you work with. The memory of mess or success when you challenge a rule can float about like a cloud the next time you are in the same situation, nudging you gently for attention. This time, though, you know from your own experience whether or when to break the rule.

At an online course on backstrap weaving, the teacher advised against using handspun yarn in the warp. She argued that handspun warp threads in a tight weave, like backstrap weaving usually is, would stick and make the shed difficult to open. I had no intention of weaving with commercial yarns, so I defied her advice and used my handspun, fully aware that I might be heading for trouble. The warp threads did cling as if their lives depended on it, and it was a true struggle to un-cling them. Tears were shed right into that shed. The weave took ages to finish, and it was not a joyful weave. The resulting textile was lovely, though (figure 6.3). I needed to find out where my limit was, how I could work with handspun warp yarn despite its tendency (or total dedication) to cling. Finding and exploring this limit helped me adapt spinning, weaving, and mending techniques to a level I could handle.

Figure 12.3. Spinning black isn't my cup of tea, but I challenged myself to buy a pitch black Rya fleece and separate the fibers. To the left are combed fibers in a worsted yarn; to the right are carded rolags in a woolen yarn together with recycled sari silk.

Breaking My Own Rules

Sometimes I challenge myself to spin something outside my comfort zone. I get to be a different spinner for a moment, step out of my regular spinning self, and discover something new. Perhaps I end up liking it. If nothing else, I will have learned and grown as a spinner.

In 2023, I bought a Rya lamb's fleece, black as the night and with 10-inch-long (25-centimeter-long) outercoat fibers—a top-quality fleece and a true beauty. I never wear black, and I find it difficult to spin black because I can't see the fibers properly. But I wanted to challenge myself and spin this fleece into a beautiful yarn. I made sure I spun in full daylight against a light background for better contrast (see chapter 11 and figure 11.2) and took my time. I ended up dividing the fleece into fiber types and spinning one strong worsted yarn from the shiny outercoat and one warm woolen yarn from the fine undercoat, with specks of recycled sari silk. I call them Panther and Starling.

Cutting Corners?

Even if I know—cognitively and through experience—why a step is important and what it does for the result, the process or a sensation can sometimes be more important to me. Before I share my thoughts on where I may cut corners, though, I want to share where I don't negotiate. Nearly half of this book is about wool preparation in one way or another. At the beginning of chapter 1, I refer to four-thousand-year-old tablets with calculations of time required for the steps from fleece to woven fabric: It took one person thirty-seven days to wash and prepare 9 pounds (4 kilograms) of fleece, and sixty-seven days to spin (on spindles). There is a reason why so much time is spent on wool preparation—it is the foundation on which the spinning and the textile creation relies. It is in the preparation steps I lay the ground for my understanding of the wool and for the quality of the yarn. Cutting corners in the wool preparation has therefore never occurred to me; it wouldn't serve me. Do try it, though, to feel and see the difference. There are, however, corners in other parts of the process that I have no problem cutting, for certain reasons.

Back to the nalbinding yarn with the blasphemous allowing of twist into the fiber supply. Out of curiosity, I started nalbinding as soon as the first ball of yarn was finished—no singles resting, no soaking, no finishing. I just plied the yarn from the two ends of a center-pull ball, skipped the skeining, and wound the yarn into a new ball straight off the plying spindle. Lots of rules were broken. However, having the sensation and process of spinning the yarn still sizzling in my fingers as I grabbed the wooden needle was more important to me in this project than making sure the twist was set and even. The approach gave me instant feedback between the steps and allowed for adjustments to accommodate to the feeling I wanted in the nalbinding process and the nalbound fabric. I created myself an overarching process that bound all the smaller processes together. Breaking the rules and cutting corners gave me an experience that stretched so much further than the nalbinding project itself. It's still hovering between the fibers in the mittens.

Frills

There are many reasons why someone spins. In some parts of the world, spinning is done to put food on the table and clothes on the back. Some spin for the results; others spin for the process. Sometimes I like some frill in my spinning, just to experience a certain sensation or way of creating. Perhaps I spin in the grease just for the way it alters the response from the fibers. My clothes may get greasy and my tools dirty. I can live with that, for the sake of the experience. Sometimes I separate undercoat and outercoat by hand, just to store the memory of the process in my hands. Holding the outercoat rats' tail in one hand and the undercoat flounce in the other, leaning my hands in opposite directions and watching the fiber types slowly glide apart into two new staples, full of possibilities, is a true bliss. Would you deny your hands that experience?

Figure 12.4. I separate this fresh off the sheep staple by holding cut and tip ends in each hand, wiggle a bit . . .

Figure 12.5. . . . and pull gently, leaving the undercoat in my cut end hand and the outercoat in my tip end hand.

CREATE: INCORPORATE

Pick a rule and spend some time exploring it. Can you feel the reason why it is a rule? Feel and incorporate the implications of the rule in your body. How much can you provoke it before it becomes a fact? How important is the rule for you? For what reason would you break it? Find what works for you.

Figure 13.1. When I spin, I can transport myself within my spinning bubble—here to the garden of the Drottningholm castle. The royal family weren't at home.

CHAPTER 13

In the Bubble

The morning sun places a soft hand on my neck where I sit at my wheel by the window. I gently turn the drive wheel with my hand and start, the familiar whoosh relaxing my shoulders. I add a carded rolag to yesterday's yarn end. I start charging the preparation with twist; six treadles to charge, make the draft, and watch the twist travel up the arm's length of yarn in the making. Eighteen treadles to add twist. As my feet work, my hands tingle with anticipation. They know what to do; they know how to keep the twist alive and open to whatever comes. They are in charge of the stretch in front of me, and they listen. Experienced thumbs roll the yarn against the twist to welcome movement, to allow the fibers to glide and conform into the thickness and the shape my hands know by now to be just right.

When I spin or prepare wool, I open the door to a parallel dimension, where only my tools, my hands, and the wool exist. I call this space the spinning bubble. In this relaxed focus, time stops, and I get to go deeper into the process of creating a yarn.

In this chapter, I want to direct your attention to the process of spinning and preparing wool and to what it can do for us in terms of a balanced and joyful experience.

The Spinning Bubble

When I spin, I allow my thoughts to come and go, without judgment, just like the fibers come from the fiber supply and go into the twist.

A process indicates that something is moving in a certain direction. It doesn't have to be in a straight line, though. My spinning journey moves with the wool as its compass, deviating from the path when it doesn't feel right. I may walk over new mossy logs and climb up on new rocks to see other views. In my spinning bubble, I see opportunities to learn, from the wool itself and from the decisions I make along the way. My brain works from the paths and symbols on the map, while my hands see the rocks and logs in the reality that is this fleece. Brain and hands work together through the wools in the past and the wool in the present, creating a path that is unique to the wool I handle. In a sense, when I spin a fleece, I spin all the fleeces I have spun before. They are there beside me, tingling in my hands and in my mind, guiding me through all they have taught me over the years. We are in deep conversation, the wool and I. That garment I wear or the cushion I lean my back against at the end of the day are so much more than textiles, so much more even than the knowledge and skills it took to make them. They are vessels of the process and the context that created them.

It wasn't always this way for me, though. There was a time when I learned how to spin, when I was a knitter who knew how to spin. But then that shift appeared, a moment in time when spinning became my main source of textile creativity, when I became first and foremost a spinner. I started to embrace spinning as a process of the mind alongside the process of turning fibers into yarn and textile (see "I Am a Spinner").

Invite the Prep

Many spinners may feel hesitant about preparing their wool with hand tools. It takes time and might not be comfortable. However, giving wool preparation the same level of attention as spinning can be just as joyful an experience.

A newly shorn fleece spread out like an early morning beach, glistening in the sun, the smell of fresh lanolin wrapped around it like a summer cloud. The moment I dip my hands into the ocean of billowing staples I enter a different realm where the lanolin has cast a spell, where time receives a different value, a tone, a new and supple dimension. In this space, I allow myself to stay and see the wool more acutely, in the same way a bee sees a flower through its compound eyes.

Figure 13.2. Spending time with the wool helps me understand it.

I pick a few staples. They part slowly, reminding me of seaweed in its muffled dance on the seabed. Fiber by fiber I tease each staple open. They spread out before me, mirroring the gentle folds of an accordion, showing me all their beauty. Occasional pieces of grass or seed emerge from the depths and slowly float to the ground, as air relaxes and opens the fiber forest. With a puff of expanded wool in one hand, I grab my card in the other and hook it onto the prickles. The fresh and untampered lanolin makes the fibers linger and the movement viscous, as if to remind me to stay in the moment and enjoy it fully. When I brush the top card onto the bottom, the wool responds, whispers its secrets to me through the slowness of the motions. My hands receive, assess, embrace this gift, and hold it gently. As I bring out a pocketed spindle, my fingers bring that gift back, in the creating of a yarn, letting the wool know it is safe in the care of my hands, that they have treasured the gift and will use it to spin its most precious yarn. The space I am invited to by the lanolin is a dimension beyond any other spinning I have known. I can't unknow it now and I am the richer for it.

As I transform the wool between my hands and my tools I get the front row tickets to my own fleece show. A shift from seeing wool preparation as a task to embracing it as an opportunity to get to know it and a privilege to spend time with it can do wonders.

As I have been writing this chapter, I have gotten stuck several times. I decided to listen to a writing pep talk by Beth Kempton about feeling stuck. She talked about energy blockages, and to allow for the energy to move to unblock. I see the same phenomenon in wool preparation—to be able to find the flow of the process, I need to unclench, detangle, and ease the grip. I need to see the wool preparation as part of the process and as deserving of the joy and flow as the spinning. Wool preparation not only prepares the wool for spinning but also prepares my body for transforming the wool characteristics into yarn. It's all connected. By finding and unblocking a channel to feeling its qualities as I prepare it, I give myself the gift of ease and flow when I spin.

A Sacred Space

The spinning bubble is a sacred space to me, a place where I can find balance, serenity, and fresh air for my thoughts. With wool in my hands, I can go to that space in my mind. My physical spinning space helps me find that sacred space. Having a nook just for spinning gives me a sense of presence. It's not just something I do passing by, it's a conscious choice I make to spend time with the wool and in the spinning. It's also a way to honor the craft and the material.

I want my spinning space to be peaceful, inviting, and with as few distractions as possible. I know that's easier said than done, especially with family or pets in the house, but a few changes can transform even just a corner toward a place of rest. My usual spinning and wool preparation space is in the living room with my spinning wheel in front of me and baskets of wool and tools beside me, marking my space. They are the physical walls of my spinning bubble. When my children were small and it was my turn to walk them to school, I used to spin while waiting for them to get ready, the morning sun shining through the window casting a warm light on my wheel. In the summer, I sit on the balcony in the afternoon shade, wind in my hair. The outdoor elements add a new dimension to the process of making, giving me a sense of being part of nature. Perhaps I take a spindle for a walk and make the path my spinning space (figure 15.1).

Spinning is always something I can return to. Sometimes I spin to find that creative spark I am looking for; sometimes I spin to understand something I find challenging. My mind is at ease here. If the world around me is in turmoil, I can rest and recharge in my spinning bubble.

CREATE: REFLECT

Now, reflect on your spinning journey so far. What is spinning to you? What does spinning do for you, apart from making yarn? Do you have a dedicated spinning space where you can enter your personal spinning bubble? Can you put words or feelings to the process and your yarn-creating bubble and what it means to you? Can you find a moment in time when you became a spinner? Take notes and read them out loud.

Figure IV.1. The pillowcase of a thousand joins.

The Map and the Pillowcase

Holding sweet locks of wool in my hands from Blanka the Dalapäls lamb, I tease her staples one by one, stroking the wool in all its length with the flicker. As the tool reaches the ends of the fibers, the staple puffs up into a cloud, ready to be spun. I place the end of my yarn on top of the preparation and flick the spindle. Within seconds, the open fibers find their way into the twist, settling into their new home in the yarn. I keep going, lock by lock, building a conical cop on the spindle in my lap, evening after evening in bed just before I go to sleep. The twirling spindle is my lullaby, rocking me to sleep.

Once the fleece basket is empty and the yarn plied, I dress my loom with a double weave in a tight sett, six hundred threads in two layers. It's my fifth weaving project altogether, my third with handspun yarn. I want to weave a tube pillowcase. The warp threads run tightly next to and on top of each other, the shed opening not more than 1 inch (3 centimeters) on a good day. I sit under the big oak tree with the loom leaning against my belly. As I open shed after shed, fibers brush against each other and loosen their grip of the twist; freed fuzz glows in the afternoon sun. Threads cling to one another; layers tangle and fuss. It doesn't take long before the first warp thread breaks. After the tenth break, my tears land softly on the fell, glistening among the fuzz.

In any other situation I would quit, and leave the project unfinished. But I had spent so much time, love, and dedication on spinning this yarn, just as I would any yarn. I owe it to myself, to all those nights of sweet sleep, to the lamb that gave me the wool, to continue, to make that pillowcase. I keep weaving. Warps break, I join end after end. As I come to the end of the warp, more than thirty threads have broken.

I sit under the big oak again, this time with a fabric in my lap. Thread by thread, I mend the joins with thrums, following the ups and downs in the weave. As I drape the fabric in my lap, I see all my mistakes—thin spots, knots, neat joins and quirky ones, and loose warps at the fold marking the upper and lower layers. I see the mistakes; I remember what went wrong, how I struggled, and how I solved the outcomes of my shortcomings. With a sudden tingle in my heart, I realize that the mistakes are there to teach me something. I expand to embrace my mistakes as a map of what I have learned.

As I write this section, the Blanka pillowcase rests in my lap. It has had its place on our couch for the past six years, supporting my back every day. The mistakes are still there, side by side with the tears, honest and open for anybody to see. I cherish each and every one of them.

Figure 14.1. When I spin on a spindle, I feel close to the process.

CHAPTER 14

Closeness

With a tuft of wool in my hand, I dress the bottom card gently, hooking the wool to the wave of bent teeth, stretching the fibers out, yawning them into full length. I brush the top card onto the wool cushion on the bottom, connecting the tools through the wool bridge between them. My hands feel the glide and resistance from the fibers and know the exact moment the last fibers glide apart and break the connection. Sometimes I linger in this moment, slow the movement down, just to feel the response from the wool. My hands aren't in the wool, yet they are connected to it through the cards. I don't have the lanolin on my skin, yet I feel through the motions how it slows the process down and makes the fibers move like honey between the cards. The carding process is in my body; my body is part of the mechanics of the carding. Through the hundreds of rolags I card from one fleece, the properties of the wool have moved into my body, fiber by fiber, to the extent that they still vibrate in my fingertips when I put the cards down and enclose my hands around a cup of tea.

• • •

In this chapter, I want you to see the benefits of working close to the wool and how much we as spinners can learn by acknowledging the power of simple tools and a straightforward connection in the spinning process. One spinning tool isn't necessarily better than the other, though; there are just different aspects of them that may work better in a certain context.

Body Mechanics

The simpler the tools, the more of the physical process is in my body. When I spin on a spindle, my body is part of the spinning process. I feel closer to the wool and ultimately to the sheep who gave it to me.

Spinning was once a vital craft for putting food on the table and clothes on our backs. Spinning on spindles took a lot of time, though. Mechanized spindles have existed in India and China for centuries. Perhaps you have seen a horizontal Indian charkha or a small hand-turned Chinese spinning wheel. This type of mechanized wheel turned up in European medieval times, heavily enlarged, as the

Figure 14.2. With the mechanics in a walking wheel, I get the benefits of speed but the disadvantages of distance to the process.

great wheel or walking wheel, but still with the same basic mechanics—a horizontal spindle operated with the help of drive wheels. The great wheels did save a lot of time, at least for spinning weft yarn. Warp yarn was usually spun on spindles only, for the high quality that was necessary. Later, the treadle wheel made its entrance and was in the great scheme of things the shortest used spinning tool for production through history.

Let's take a closer look at a treadle wheel and compare it with a hand spindle. On the one hand, we have a mechanized piece of furniture with wheels and treadles on which the spinner can create yarn through the connection of a drive band. The spinner can adjust tension, speed, and uptake by turning knobs and changing pulleys. On the other hand, a hand spindle is a stick and the weight of a whorl. How is it, then, that a spinner can create the same result with these two remarkably different tools? The wheel is mechanized to save time and effort. But if we can get the same result from a wheel and a spindle, where, then, are the mechanics of the spindle? Well, some mechanics are in the spindle, of course: weight, size, whorl construction, and whorl placement are factors that, when in motion, can have an impact on yarn thickness, speed, spin duration, and technique. But when we spin on spindles, we as spinners are part of the spinning mechanics. Read that again: we as spinners are part of the spinning mechanics. When we spin on spindles, we control with our bodies that which has been lifted out and placed in the mechanics of a spinning wheel. When we spin on spindles, with the mechanics in our bodies, we can feel the spinning process physically and mechanically in our hands and bodies. With this in mind, I am convinced that it is easier for most people to learn how to spin and understand spinning mechanics on a spindle than on a spinning wheel.

Muscle Memory and Connection

When I was little, I used to look through a popular science book, often at one picture that particularly tickled my fancy, a man stretching his hands toward the reader, tongue out. The hands were gigantic in relation to the rest of the body, as were the tongue and lips. The image showed the cortical homunculus, something that is nowadays represented by two 3D models: one motor homunculus and one sensory homunculus, showing the levels of sensory and motor representation in different areas of the body.

Figure 14.3. When I spin on a spindle, I have the yarn right in front of me, always ready for quality control. Here I slack the yarn to check the twist.

The proportionally larger something is in the model, the more motor and sensory representation it has in the brain. Looking at those gigantic hands makes me realize that working with the wool in our hands and with hand tools gives our brains more information than when we use more mechanized tools.

Let's take carding with hand cards as an example. Every time the cards brush the wool, I feel the motions through my hands holding the handles. When the fibers glide past each other, I feel how they glide—the pace, the resistance, the smoothness. I can also get a sense of how the elasticity works—how the fibers stretch in the carding strokes and how they bounce back into their natural crimp when the fibers come apart. Perhaps there are tangles in the fibers that my hands register as the resistance that breaks the rhythm. When the fibers glide apart and separate between the cards, I also feel the length of the fibers (figure 8.1). Finally, I can get a sense of whether I have dressed the cards with an appropriate amount of wool depending on the feeling of the top card against the bottom. All of this is information I need to understand what the wool wants and to make informed decisions on tools and techniques in the upcoming steps in the process. A drum carder would no doubt get the job done more quickly, but the information I get from hand carding wouldn't be available to me.

With the close physical connection to the tools and the wool, I also have them close enough to see the process. When I card or comb, I see tangles or wool characteristics as I prepare it, and when I spin on spindles, I see a bite-size portion of the yarn right in front of me (figure 15.4). Through the combination of the visual and the physical experience of the process, I believe I can understand more about the process and the wool, and how I need to respond to what I learn. We will dive deeper into these aspects in chapter 19.

Simultaneous and Linear

The spinning process is simultaneous on the spinning wheel—the twist enters the yarn while I handle the fiber. At the same time, I operate the mechanics with the treadling of my feet to add twist to the yarn and feed it onto the bobbin. Wheel spinning is also causal—when I treadle, I can't separate one action from another. Spindle spinning can be a simultaneous process, too, but a lot of it is inside the body of the spinner. A beginner spinner can find the simultaneity and the speed of a spinning wheel overwhelming and difficult to grasp. A spindle spinner, however, has the choice to break down the simultaneous parts into linear units and absorb them one by one. This is what happens when we spin with a park and draft technique—we hold the spindle still or place it between our knees as we draft the fibers, set the spindle in motion without drafting, and then stop it again to draft. Of course it is possible to treadle the wheel to add twist to the yarn and then stop to make the draft, but without the precision of spindle spinning.

With the closer connection to the wool and the physical spinning process, I can make more informed decisions. When I know how the fibers work in movement and stillness, I can be more subtle in my hand movements to adapt to the wool and the tools in my hands. I believe that this understanding creates more space for me to go deeper into the spinning process and enjoy it more.

Spindles as Resource and Joy

Don't get me wrong. I do use my spinning wheel a lot, but I don't abandon my spindles in thinking the wheel is better or because it will save time. I use spindles for the sake of spinning on spindles. It doesn't have to be more complicated than that. Feeling both wood and wool underneath my fingertips gives me a sense of grounding and connection to the spinning process. The simplicity of the mechanics makes the spindle an easy tool to bring anywhere. I love walking and spinning with spindles in the summer. There is something about the movement forward in space that also seems to bring my thought and learning processes forward (figure 15.1).

When I teach spinning, I teach on spindles, but I also learn best myself on spindles. They are my firsthand choice when I want to explore a new technique or just experiment and sample to find the yarn that is the best for the wool I'm working with (see chapter 6). What I have developed on a spindle I can later translate into wheel spinning if I choose.

When Hand Tools Don't Match You

Working with the hands and with hand tools is all good if you have the strength and dexterity to do so, but what if you don't? How can you adapt tools and approach the wool when you, for different reasons, can't work through a whole fleece with hand tools? If you can't process a whole fleece with your own mechanics, perhaps you can explore the wool with hands or hand tools to find out what it wants or prepare wool for one skein with cards or combs. The same goes for the spinning part. If spindle spinning works for you but not on a larger project, you can explore a fleece or a technique with a spindle on a smaller scale. It can be a good way to get to know the wool with less pain or strain. With the experience of exploration or processing of the first skein in your hands, you can translate that into more mechanized tools. If you can't work with hand tools at all, you can spend some more time with your hands in the wool at the picking stage and explore it more thoroughly there. This way you can still get the feeling of the wool in your body and an understanding of the wool characteristics. From there you can make informed decisions with less pain and strain.

EXPAND: CONNECT

List the connections you have with and through the wool, be it a sheep owner you know, a wool broker you got the wool from, pastures you pass daily, having cuddled the sheep itself or another member of the flock, the connections of the hands in the fleece, the fleece between your hands and tools, the spinning process in your body.

When you look at your list, can you track the connection from the yarn all the way back to the sheep that gave it to you? What does that do for your experience of the process from wool to yarn and of the yarn you have spun?

Figure 15.1. The slowness of spindle spinning is an advantage; it's time I get to spend with the wool and the technique. Here I'm using an Andean pushka.

CHAPTER 15

Slow

With a spindle in my pocket, I go outside, fill my lungs with morning air, and start walking. I watch the landscape passing slowly, note every tree, every moss-covered stone, and every meadow. I pick up my spindle and give it a gentle roll, flat hand against my thigh. With a sigh of relief, the spindle starts twirling below my hands as I tend to the wool. I see it all right in front of me—the rolag in one hand, the yarn in the other, and the beating of the spinning heart moving unhurriedly between spun and unspun. Even if I were to close my eyes, I would see it, through what the fibers tell me if I care to listen. A subtle roll of the yarn against the twist, and I open up to a flow I can steer toward fibers or yarn from the tingle of the wool against my fingertips. In this close connection between hands, wool, and senses, I don't even need to think; I have enough information without dipping my thoughts into the pot as well. My follower hands listen to the cues from the leader wool, constantly paying attention to where the wool steers me. The wind in my hair is our music, and the path I walk is our dance floor as we move forward, deeper into the dance of making. Through the slowness of making yarn on a stick I feel stronger, shine brighter, and learn deeper. I'm not in it for the speed.

• • •

In this chapter, I want you to learn what we can gain by spending time in the spinning—how to make informed decisions on what steps to take, skip, or linger in. Even if we choose not to spend time in a stage, just learning about the step in slowness can help us make that decision. Slowness is to me one of the superpowers of spindles.

Go Slow to Go Deep

The Centro de Textiles Tradicionales del Cusco (CTTC) is a nonprofit organization in Peru.[1] It was founded in 1996 to work with communities from the Cusco region to revive textile traditions and empower weavers, especially women. Nilda Callañaupa Alvarez is one of the founders of the organization. The work began in the 1970s when Nilda and a group of women started to spin and weave together. The traditions of spinning, dyeing, and weaving intricately patterned textiles in the region were beginning to disappear, and the group recovered traditional designs and natural dye processes and relearned techniques that had nearly been forgotten. Children in the region learn to spin on hand spindles, pushkas (figure 15.1), at an early age, often in the pasture as they tend to the sheep and alpacas. Later they spin a large part of the yarn needed for their families' weaving projects. The fiber is teased by hand into continuous rovings and then wound around the wrist to spin from.

Nilda started her spinning journey with the pasture as her classroom, too, and is today the author of several books about spinning, weaving, and dyeing in the Andes. I asked Nilda, who has little experience with wheel spinning, what spindle spinning means to her. For her, the spindle is a continuous company she can bring anywhere she goes to produce any thickness of yarn she wants. She doesn't enjoy times when she is not spinning, and she sees spinning as a way to wind down after a long day. It brings her back to her time as a young shepherdess.

Speed is a power that will give us the finished product faster and the joy of having something made. Slowness, on the other hand, can offer us time to spend in the process, in the making. This is why I spin. I love using and wearing things I made, but the time I spend in the process from fluff to stuff is a state of mind where I find joy. And the yarns and textiles that come out of the process of making are reminders of that time.

1. To donate to the CTTC and their causes, see https://store.textilescusco.org/donate/.

Figure 15.2. When I work with my hands and hand tools, I gain lots of knowledge that I wouldn't get with more mechanized tools.

Slowness, whether it is in the manual preparation or spinning with hand tools, is time I get to spend with the wool; it is time to understand the technique and process the characteristics of the wool in my mind. It is time to dig into my bank of previous experiences and weave them together with new additions from the wool in front of me. A friend in a period of fragility got the advice to strive inward rather than upward. This has been a guide for me since she shared it, and it certainly applies to my relationship to spinning.

Spindle spinning can be a reminder and connector back to the first spinners, to all the fabric spun on spindles around the world through millennia. It is a chance to revisit the Bronze Age era wool economy when spinning was the backbone of whole societies (see chapter 1), and still is in some communities.

Dissecting for Understanding

Speed can no doubt get things done faster. The mechanization of spinning and the speed of production that followed was the start of the Industrial Revolution that has changed whole societies. To me as a hobby spinner, these advances can sometimes take something away. Every second I get to spend with the wool, in preparation or spinning, every slow moment where I get the chance to grasp what is happening, is time I gain in knowledge. There are no shortcuts to understanding the wool and how it behaves. When I spin on spindles, my hands and the heart of the

Figure 15.3. Slowness is a superpower, and I'm not in it for the speed. My egg shelf spindle rack reminds me every day. For spindle makers, see appendix 1.

Figure 15.4. Most spindle types allow the spinner to have the heart of the spinning right in front of their eyes for section-by-section quality control.

spinning are right in front of me. In bite-size sections I get the time to survey and understand the wool and adapt to how it works in motion. I get the time to do quality control as I'm spinning.

We discussed the possibility of dissecting the mechanics of spinning into separate and linear parts in chapter 14. In a park and draft technique where I single out the drafting part, I get the time to explore it separately: In what circumstances do I get access to the point of twist engagement? What needs to happen for the fibers to open up enough to unlock themselves without coming apart completely (see chapter 9)? How wide is that window? How much do I need to roll the yarn between my thumb and my index finger to find a smooth pace? Or, at what point do I let go of the yarn with my spindle hand to let the twist into the drafted fibers? When I have found ease and confidence in the drafting I can gradually unpark the spindle and work toward simultaneous spinning.

In a double drafting technique, where I first spin one arm's length loosely, make adjustments in smaller portions, and then add the final twist, I get to spend more time with each section, while at the same time I can spin a more even yarn.

In wheel spinning, the tension is set on the wheel. With any kind of spindle spinning, I have more control of the tension:

- In a suspended spindle, the weight of the spindle sets the tension, and I feel it in my spinning hand (figure 15.5). The tension between my hands mirrors the tension from the spindle suspended at the other end.
- On a floor supported spindle, the tension can be between my spindle hand that holds the shaft and my fiber hand. If I hold my spindle-hand thumb on the yarn, the tension is more directly between my hands (figure 9.6).
- In supported-spindle spinning, and to some degree in-hand spindle spinning, the tension is fully between my hands, giving me the opportunity to be in total control of the tension. I can spin a very fine yarn on a supported spindle through that control.
- A horizontal spindle can give me information about the tension in two steps, first in the horizontal stage between my spindle hand and my fiber hand, and then in the suspended part of the process between my fiber hand and the weight of the spindle.

In any spindle-spinning project, my hands learn what the tension should feel like throughout the spinning for one skein and, for a larger project, even between skeins. Through the tension control in combination with the opportunity to slack the yarn for a ply-back (figure 14.3), I can keep regular control of the twist without measuring angles. With the pull of the ends in the ply-back, my hands recognize the response from the yarn and learn to keep an even

Figure 15.5. A suspended spindle in all its simplicity.

twist through the information my hands get from the yarn. With the deeper understanding of the spinning process through spindle spinning and the mechanics in my body, I believe my hands can work more by touch and with more ease than on a spinning wheel. I don't have to pinch because I don't understand what's happening. Instead, I can spin more consciously and be in the spinning state of mind.

Time Gained through Slowness

The time it takes me to pick my wool by hand, to card my wool with hand cards, or spin it on a spindle, is time I gain. Sometimes that time is what I need to understand something. In another context, I can use what I learn in slowness for a faster technique. The possibility of spinning anytime and anywhere with a spindle, is that slow?

My why of processing and spinning with hand tools is about the process—of physical and mental making and of learning and reflecting. My tools and techniques may be slow, but to me spinning is also therapy, balance, and peace of mind. I'm not sure that can be measured in time. To me, spinning feels quite close to meditation. Slowness is a superpower that allows me to go deep and understand the process physically, cognitively, sensorily, and spiritually. In her book *Secrets of Spinning, Weaving, and Knitting in the Peruvian Highlands*, Nilda Callañaupa Alvarez describes spindle spinning as "a community-wide language that everyone understands," as well as "a personal language that individual spinners use as they sit for hours in personal contemplation." That personal language and contemplation is often worth much more to me than the speed of producing a yarn.

EXPAND: FEEL

Choose a technique that is slower than your go-to technique. If you usually spin on a spinning wheel, pick up a spindle; if you use a drum carder, card with hand cards; or if you tease with combs, use your hands to tease. Now, single out one part of the process to focus your energies. Perhaps you park your spindle and focus on the drafting or the point of twist engagement. You might explore the rhythm of hand carding or close your eyes as you tease staples by hand. Make a few notes if you want to, pick a new technique, or explore something different. You can do this practice many times and at many levels.

CHAPTER 16

We Need to Talk

Through years of blogging and teaching spinning, I have marveled at the power of words—of how they can help me articulate and chisel out a feeling or an idea. I need to choose my words carefully for the reader or student to be able to follow my train of thought and still have space to ponder on their own. Through crafting my words, I understand myself better and take the concept to a deeper level, and yet another level, and so on, deepening my understanding of whatever is on my mind. There is a feedback loop and an inner conversation here, too, that I need to keep going to be able to sharpen my choice of words. I see the same process when we learn to spin—by articulating what we do we can understand better. With relevant vocabulary for techniques and processes, we give ourselves verbal tools for exploring these further. By keeping the conversation going, we can learn to listen to what our bodies and minds are telling us as well.

In this chapter, I want you to learn to move your understanding forward and deeper by articulating what you do and listening to the reply you get—from people around you and from yourself in your inner conversation.

Understanding through Expression

We need to talk about what we are doing, using relevant spinning vocabulary and making it present and visible. We need to lay out the words we have established, for easier access to understanding what we do, why we do it, and what implications the decisions we make have for our spinning process.

I'm in the classroom teaching my annual course "A Spindle a Day." In five days my five students will

Figure 16.1. Narrative spinning is an exercise where two spinners help each other express what they are doing and through that understand more about technique and process.

learn how to prepare wool and spin on four different spindle types. Each day they will build on the skills and the knowledge they acquired the day before. We talk. I introduce terms for techniques, tools, and parts of the process. I repeat the terms every time I return to the concepts they stand for, and the words slowly but surely become part of the students' vocabularies. When we dive deep into the Twist Model (see chapter 9), we use the terms I have established for the model: the point of twist engagement; unstable, stable, and semi-stable; and opening up the twist. I refer to the section between my hands as the heart of the spinning and use the phrase in connection to the Twist Model as well as in other contexts.

By using the terms, we create a common spinning language, a scene where the spinning is performed and to which we can refer in the intermission. We talk and we listen. Whenever we get back to the heart of the spinning or opening up the twist, we all know what we are talking about and what part of the process we need to locate in our minds and unfold in a new setting.

Curiosity

By now the classroom is sizzling with energy. The students bubble with questions, reflections, and inspiration. There is an ongoing conversation. I tell the story of my nalbound mittens where I cut lots of corners for the sake of the immediate feedback from spinning to nalbinding (see chapter 12). One student gets inspired and starts to knit from what she has just spun, just to experience the feedback in the knitting from the choices she made in the spinning, and takes the resulting insights back to it for the next skein.

In my teaching, I want my students to understand what they are doing and why something works or doesn't. I want them to take their understanding forward. I am that annoying teacher who insists on asking why the student thinks a particular result happened, what they can change, and how it can feel different. I encourage them to take notes, reflect, and talk about what they experience, to understand themselves better and share with the other students for a deeper collective understanding. We need to keep the conversation going.

I have two complete beginners in the class, and they are exhausted after the first day of introduction, wool knowledge, wool preparation, suspended spindles, and the Twist Model. Still, on day 2 they recognize the terms and seem to be more comfortable walking around in the spindle forest without getting lost. Even if they do get lost, it is temporary; they know they can find their way back to the track by holding on to a tree and taking a breath. They learn to spin on the floor spindle and receive new terms, like *changing angles*, *long draw*, and *upper and lower cops*. At the same time, we get back to and build on what we talked about on day 1.

When I reintroduce a term from yesterday's spindle type, the students understand what we talk about in today's, even though we haven't learned the technique for it yet. At the end of day 2, one student says, "After the floor spindle I understand the draft in the suspended spindle better!" When we get to the in-hand (grasped) spindles on day three, the students have both technique and terminology. All they need to add is the grip and, if they are up to it, the distaff. Even if they struggle with keeping the spindle moving, they know and use the terms with ease; they can express their struggles and successes. One student sees the pattern of changing angles and bursts out, "The in-hand spindle is really the same technique as the floor spindle!" The students transfer the knowledge between spindle types, using the vocabulary we have built.

Narrative Spinning

On the fourth day, we do an exercise I call narrative spinning (figure 16.1). We use the fourth spindle type of the course, the supported spindle. The name of the exercise comes from when I was practicing driving many years ago. My driving teacher asked me to narrate my driving—to say out loud what I saw, how I needed to respond to what I saw, and what the implications were to that response. For example, "I see a 20 miles per hour (30 kilometers per hour) sign ahead. I need to engine brake to slow down to the speed limit before I pass the sign. There is a reason for a 20 limit, so I need to watch for pedestrians, crossings or bumps in the road." In narrative spinning, one student, the narrator, spins and talks another spinner, the listener, through what they do: "I set the spindle in motion, move my spindle hand to the yarn and the yarn angle to almost 180 degrees from the spindle. I roll the yarn against the twist to find the point of twist engagement. Meanwhile I hold the rolag with a gentle grip and make sure there isn't too much twist between my hands. I want to keep the twist open and the section between my hands

Figure 16.2. We talk a lot about the process in the classroom. In the circle we are equal, and every student can see and hear each other and take part in individual feedback.

semi-stable until I am happy with the thickness and evenness. After that I can add the final twist for a stable yarn and slack it to check that the ply-back matches my earlier ones. I change the angle of the yarn and roll it onto the upper cop." The listener may ask questions like, "How do you determine the twist in the ply-back?" The narrator may reply, "Well, I look at it to see if it matches the earlier ones in twist angle and thickness. I also feel the tug in the ply-back between my hands. An aggressive tug will indicate quite a high twist. I compare it to the memory of earlier ones and make a judgment about its likeness to my model."

After about ten minutes, the students switch so that the narrator becomes the listener. The conversation in the classroom moves effortlessly, and the students develop their capacity to narrate the spinning together, as a joint effort and a deepening of their understanding of what they do. By articulating the spinning for each other, they automatically make the process clearer to themselves. Many of the students discover new aspects of what they do the moment they build a new sentence. Furthermore, they get a glimpse of each other's perspectives as narrators and listeners. Through the conversation, they explore the spinning and move their joint learning forward together.

Mind, Body, and Spirit

We keep diving deeper into techniques—talking, listening, and reflecting. Meanwhile, our bodies are talking to us, and we can choose to listen to them. Intuition and sensory and emotional response to the situation can tell us what feels comfortable or not. It can be subtle things like air quality, noise, or light, but also more physical signals from positioning and posture. The body knows more than we may be aware of cognitively, and we need to listen to it and adjust.

The students analyze and reflect before they ask questions or think about what they "should" do. They now understand what opening up the twist does for the ease of the spinning, what the teasing before carding does for the quality of the finished yarn, and what the evenness of the rolags does for the spinning flow. They also reflect on what is appropriate for their own bodies. The student who is living with the implications of a stroke realizes the floor spindle is what she can handle at the moment. She struggles with the supported spindle but knows she will become more comfortable with it over time. The student who has an illness she compares to a chronic massive burnout realizes she can't card without pain. She decides not to go forward with the floor spindle; she knows she won't be able to card the rolags for it. The supported spindle with commercially prepared fiber is where she finds her ease. Through the technique she learns in the course, she manages to load a lot more yarn onto the cop without pain than she could before. The student who bought 11 pounds (5 kilograms) of Icelandic fleece that is compacted and tangled experiments through class and in the evenings and finds a way to process the wool that works for her and the tools she has. She wants to work with a small number of tools that suit her needs and realizes with a sigh of satisfaction that the limitation is the freedom.

When we see the body as a learning site—embodied learning—we can make decisions to move the learning forward with ease and comfort. When I see grimacing or signs of pain or blockage, I decide to break class with a movement series (see chapter 11) to release the block.

The organization in the classroom is part of our joint learning. We sit in a circle, like sheep in a flock, with an empty floor in the center and the tools on tables along the walls behind us. Nobody sits behind or in front of anybody. It is my impression that the group feels safe in the classroom. Students and teacher are all equal parts of the circle, and we can see and hear each other clearly. If I give individual feedback to one student, the others are still close enough to take in our conversation and learn. As a teacher I can see everybody's postures, tensions, and facial expressions and ask them what is on their mind or what they can do to spin more comfortably.

I start every day by asking the students how they feel after the day before and end every day with a few minutes of quiet contemplation and an opportunity to make some notes. Anyone who wants to share their reflections does so, and I find we often wrap up the day in an open and respectful discussion on a spiritual level—how spinning and sensing the wool in our hands make us feel or what spinning does for us mentally and spiritually. Several of the students refer to what they have learned from tools, wool, or the other students. This is a vital part of the course. We are physical and spiritual beings, and these parts of us can't be lifted out of the classroom, especially when we learn something physical. We need to talk about being in our hands and minds. We need to talk to be able to listen.

A Taste of Wool and Words

On the last day, I invite the students to a wool tasting. They get to explore five new wools for fifteen minutes each. They all get one mystery wool at a time and do what they like with it. Prepare, spin, and sample if they like—all individually and in silence. They get a chart with a few guiding questions, and they fill in whatever they like; perhaps they add a sample to the chart. Everything is for them—they decide what to do with the wools, what tools to use, and what to note on the charts. When the course is over, they get to keep the charts as diplomas.

The exercise is a chance for the students to get acquainted with the five new wools and make decisions about what to do with them. It's not about achievement; nothing is right or wrong here. It's just a chance to explore the wools and the skills they have acquired during the week. The dialogue from the spinning narration has turned into an inner dialogue with the wool. They survey their wools, listen to them, and make decisions from wool characteristics, tools, and their personal skills and preferences. They know now what they are doing; they are confident in their skill levels. Even if they may have started as beginners, they know the theory, they have learned through their mistakes and their experience and through having the words, and they have used all of these in companionship with spindles, tools, and wool.

For the seventy-five minutes the students are silently tasting their wools, I watch them. I know they are listening to the wool. My heart smiles with joy and pride at what they have learned and are now practicing on their own. I see their gazes steadily on the hearts of the spinning, in deep contemplation and conversation with the wool. When the wool tasting is over, I know something has shifted. There is something sweet in the air, a softness tingling in the students' eyes and hands.

The last thing we do before we all head home is a spinning meditation. I guide us through the session and touch on the sensation of the wool and the spindle in our hands, the roles of the hands, and the motions of the body. The students get the chance to experience the spinning process deeper in body and spirit and to be in the making. Toward the end of the meditation, I invite them to close their eyes and practice their trust in themselves to listen to the wool. And they do. At the same course in 2020, a beginner spinner said after the end of the meditation that the piece she had spun with her eyes closed was the best she had spun all week. Another that she would put together a basket for spinning just for the joy of spinning and not for results. The joy in the spinning matters. We all exit the circle and go home a little lighter, a little more joyful.

EXPAND: NARRATE

Now, try narrating your spinning (or wool preparation). You can do this in several ways. If you have a spinning friend, you can follow the spinning narration described in this chapter. Do it for around ten minutes before you switch roles. You can do it with someone who isn't a spinner, too, perhaps with the exception of the role switch. This partner will probably ask different questions than a spinner would. It can actually be a good thing—narrating spinning to someone who doesn't know what is happening and with terms you'll need to explain might challenge you more than narrating to a fellow spinner. A third option is to narrate your spinning to yourself and record the narration. Listen to the recording and make notes of what you find. Narrate again, record, and see what comes.

As a bonus exercise, you can narrate on an embodied and spiritual level, for yourself or to a friend. Narrate what your body tells you: how you feel the fibers gliding past each other as you open up the twist, what your arm tells you as you raise it and what makes you decide when to roll the yarn onto the cop, what you feel during the spinning process, and how you are in your mind and your hands as you spin. There are many avenues to explore here.

Figure 17.1. My first handspun yarn for two-end (aka twined) knitting was far too loosely spun and needed mending. (Pattern by Berit Westman.)

CHAPTER 17

Embrace Your Mistakes

Choosing to expand from the mistakes we make and seeing the progress from where we began can contribute to a more joyful experience. In this chapter, I want you to learn to see your mistakes as guides to understanding the whole in the small part you are working on.

Really?

It may seem odd, but this chapter is a pep talk. As spinners, we will have setbacks somewhere along the line, and we will make mistakes. While failure and mistakes may feel disheartening, I see them as gifts. I get the chance to reflect on what happened and to explore how I can do it differently. We fall in order to learn how to get up; we make mistakes in order to learn how to mend them and move forward. By going back to the possible source of the setback, we get the chance to learn—physically and intellectually—why it happened and what we can do to deal with it if (when) it happens again.

Being Ready

Before I feel the consequences of a rule in my hands, I rarely understand the implications of established rules. The mistakes help me understand them on a physical as well as an intellectual level. Usually, I discover my shortcomings from preparation or spinning several stages later, in the textile I am creating.

In a course, I guide a student into making larger combing movements with longer fibers to avoid loops that tangle (see chapter 8). She nods politely and increases the distance between the combs. Hours later, when she is combing her 12-inch-long (30-centimeter-long) Icelandic outercoat she bursts, "Oh, now I see why I need to make larger movements with longer wool!" The need to feel in our hands what we read or hear about becomes obvious. I had served the information to her before she was ready to receive it. She had understood what I had said but wasn't in the same place I was. When she experienced the loops herself, she remembered

Figure 17.2. Loops and tangles can be a challenge if we don't make the circles large enough when we comb long fibers.

and realized on a physical and technical level what they meant for her current situation. Talk about feedback loops!

I remember a passage in Robin Wall Kimmerer's book *Gathering Moss* about learning in indigenous communities, that knowledge can't be taken but instead must be given. We can ask a question and get an answer, but we also need to be able to take in the answer from where we stand. I may understand the words, but not the physical or technical implications of them. By watching, listening, and experiencing, we learn from our context and capabilities. The knowledge will unfold when we are ready to receive it. There is no fix-it solution to all wools and all tools. We need to learn how to listen to the wool we have in our hands and experience how it works. By going back to the roots—to the characteristics of the wool, the tools we have, the way we learn, and the experience we have—we can become ready to move forward. The characteristics and movement patterns of the wool are the foundation from which we make our decisions. It can be different decisions for different people, contexts, and situations, but it is in that foundation we need to start exploring to expand. It is when I take the time to experiment, explore, and listen that I find my way home to the joyfulness of spinning.

Go Wrong to Find the Right Way

When I teach, I mix theory with practice, connect back and forward. In the classroom, we weave it all together with the experience we have from before and build on it in the course. Any mistakes that happen are perfect opportunities to be curious and learn. I pick up the mistakes and reflect with my students from there. We look at why something happened, what we can do to mend the mistake or to avoid it in the future. With practice, we can internalize patterns and movements, but mistakes do happen, and we need to deal with them. Beth Kempton writes in her book *Wabi Sabi* about failing our way forward. What we perceive as a failure doesn't have to be the end of the story, but the beginning of the next chapter. This can only happen if we accept imperfection and choose to move forward. We are all learning, all the time, and we can choose to learn from our mistakes as students of life. If we take a step back and

Figure 17.3. With the wool dressed across the whole width of the carding cloth there is a risk that the outermost wool isn't carded at all. The rolag stretches long and becomes uneven and thin. This will make the spinning more challenging and the yarn uneven. Compare with the rolag in figure 8.10.

investigate our mistakes, we can learn what they can teach us and how we can expand from there.

I usually have a mixture of beginners and experienced spinners in the classroom. Some of the experienced spinners can have a hard time accepting that they are in fact beginners in the technique they are there to learn. They are impatient and surprised when they don't grasp the new technique after a couple of minutes. I remind them that they have a lot of experience, but not in this way of spinning. They need to give themselves and their bodies time to be in the new movement pattern and the new mode of spinning.

Beth Kempton writes further about failure and that to define what failure is we need to define what success means to us. Is it a measurable goal, or is it a feeling? We also need to decide what to do with the failure once it hits us in the face. We can't expect to reach perfection, but we can expand from where we stand. We can look at our first twenty wonky rolags (see the "Play: Shape" exercise in chapter 8) and choose to see the progression from the first to the twentieth. We can choose to feel the joy of making with rolags that support the flow of spinning. A perfect example of embracing our mistakes is visible mending—a way to honor a garment by making the mending visible and beautiful instead of hiding what has finally withered around our knees and elbows after years of service.

We can choose to be curious about our mistakes. There may be several doors to open to find the way forward for you and for the situation. Open them and peek inside. Some may be right and others not. By opening the wrong doors, you may learn why they are wrong in this context. Play with your options, find your edge, and learn from the experience. Later you may remember the doors that were wrong then and realize they are the right ones in another context. Mistakes are part of the learning process and will take us further along the spinning journey. Embrace those wrong doors!

A Map of What I Have Learned

The thing with learning a craft is that you see the result as an open book—all the wonky joins, the thin spots and the slubs, over-twisted and under-spun. Every time I look at a wool preparation, yarn, or textile, I see my mistakes. And I see them as a map of what I have learned and how I can move forward.

I still struggle with failing rolag joins like those in the pillowcase in the introduction to this section (page 113). When I make joins, I think I am so thorough, but in a warp, my join-making shortcomings are revealed. I watch the warp thread getting thinner for every new shed and feel the agony rising when the final fibers fizzle out beyond their grip and break the orderly structure of stretched threads. My handspun warp yarns still break, especially if I use singles yarns, but the breaks are fewer now. I love weaving. I still see myself as a beginner, though, and there are many rules in weaving that I blissfully don't know. Spinning is always my foundation, and I explore weaving from there. If that means fuzzy and breaking individual warp threads with questionable joins, I just need to deal with them.

I knit my first two-end knitted mittens from my handspun yarn back in 2012 (figure 17.1). They now have bright mendings on the thumbs. I spun the yarn in way too short and fine fibers with too little twist (that untwists even more in the two-end knitting technique), and the yarn broke several times as I knit them. I fulled them heavily, but they still got worn through at the thumbs. Since then, I know I need to choose both fibers and technique extra carefully when I want to create a yarn for two-end knitting. I am grateful that my mittens showed me how, and for the mendings that remind me. Several years later I made a pattern for half-mitts with handspun Z-plied yarn with a mix of long and short fibers (figure 5.1). You can find it in the fall 2019 issue of *Spin-Off* magazine.

Mistakes may be a more brutal way to learn than just having the experience in my hands, but they also give me a sharper response. Usually, things don't have to go this far for me to learn and avoid mistakes, but when they do I see them as gifts. Every decision and every mistake I make is vital for my learning process. That first wonky rolag may look like just a tuft of wool, but it is the first on a long journey of rolags that carry the memory of that first one with it. What has once been carded can't be uncarded; what has once been spun can't be unspun. The experience is woven into my hands and into the map of what I have learned.

Learn How to Fix Them

I make lots of mistakes, and I know I will be making them again. I read up on how to make better joins, and I have improved. Still, I will continue to see joins breaking in my warp. Rather than avoiding weaving with my handspun yarn, I learn how to prepare the wool for a stronger yarn and how to mend the weave. After all, I did learn to weave to be able to spin yarns suitable for more than just knitting. I explore my limits to learn what degree of fuzzing and breakage I can tolerate. I am now a master of mending broken warp threads. From several two-end knitting projects I have a better idea of what fibers, preparation, and twist to use for two-end knitting yarn.

In looking for the cause of a setback or mistake in the previous step(s), I can improve while at the same time learning why that step is important. I may decide to sample and swatch to find the yarn I am looking for, or I can go with the flow and just be in the spinning. I may stick to an idea that is less than ideal for a project, but with that knowledge and with the reasons I have for sticking to my idea, I can be aware of the risk of failure and plan how to deal with it. So go ahead, explore to expand, make the mistakes, and thank them for being your teachers.

EXPAND: NAVIGATE

Pick a mistake you have made, be it a wonky rolag, an uneven yarn, biased knitting, worn out mitten thumbs, or broken warp threads. Investigate it. Try to find what you could have done differently. Check the troubleshooting lists in chapter 8, go through the Twist Model, seek out tutorials for making lasting joins in different spinning styles (I will). Whatever mistake you choose, make sure to go a couple of steps back in the process to find where you can do things differently. Play with it; make some samples where you push your limits to find a way that works for you. Learn how to fix the mistake when it happens again. Then look at your chosen mistake and thank it for showing you your way forward.

The Wool Spirit

I am the wool spirit. I am the breath and the soul of the wool, the air between the fibers. I make sure the sheep is protected from the elements. I make the outercoat strong and resistant, bold and buoyant, the undercoat light and airy, embracing and kind. I bind the two together, bundled in staples, allowing them to work in unison for the sheep.

I know you want all the answers to how to spin me, a quick fix to the yarn. I know all the secrets of my master, yet I only tell you when you are ready to listen. If you are to destroy and re-create, you'd better do it with respect for what was once created. If you ask without listening, you will find nothing. I will give you the clues only when you deserve them. The wool is a gift from the sheep. You need to earn it and give your dedication in return. You see, you are in my hands now. You need my guidance. You will be in my hands until you let go of the result and trust yourself in your own hands.

When you lighten your grip, I will sing the fibers to you. When your hands are ready to listen, your senses will be clear enough to hear my song. When you allow your hands to ask questions, softly, only then do I sigh signs for you to ponder upon. Only then can you deepen your understanding of the works and the wonders of the wool. My voice will be soft like a breath, whispering my secrets to you.

The more time you spend in the wool, the deeper you will learn. When you are retreated in your bubble, sensing the flow of the fibers in the rhythm of making; when you are in your spinning zone, in your hands and in creating; we meet, in our joint celebration of the wool. I will be there with you. When you listen to the wool, I know you are grateful for the gifts you receive.

Long after you have put your tools down, you will feel the vibration of the wool in your hands and in your heart, echoing my whisper. When you dress in wool created by listening, I will protect you, too. ✻

Figure V.1. Is this what a wool spirit might look like?

Figure 18.1. We get a lot of information from all the senses, especially through touch, and even more with the input of a combination of senses.

CHAPTER 18

The Knowledge of the Hands

The moment the spindle starts
dancing the wool
through my hands,
hormones
burst open all doors,
sinking shoulders, heart rate and pulse
into calmness and blissful joy.
Mind and body have entered
the spinning zone.
Fingers eager to make sense of sensations,
just like when they were chubby and new
and explored the world
through bright and brightly shining
freshly made skin.
Lightly at first,
sensing the structure,
then deeper,
kneading every detail
into every nook and cranny
of the ridges of the fingerprints,
exploring from every angle,
assisted by what the eyes see.
Hands magnifying the sensations,
passing on to the brain
what they have learned,
brain painting a picture
of what touch and sight have absorbed,
sending one image back to the hands,
one to the bank of experience.
Hands, eyes and brain
deeply focused
on the heart of the spinning,
between open fibers
and structured yarn,
a kaleidoscope of knowledge
from their collective findings.
With the sensemaking from the brain,
the hands grow their knowledge
of the wool,
adding what they feel to
earlier journeys,
considering
what the length of the fibers means for preparation,
what glide means for grip, lanolin for drafting,
elasticity for textile structure,
fineness for intended use.
Fiber by fiber,
step by step,
from fleece to fabric,
understanding the wool anew
in its every new stage.
Even with tools
between hands and wool,
the tentacles of the hands
feel the response from the fibers
as though the tools
were the hands themselves.
There is a trust in the hands
to register the information
in the brain to interpret,
to place it in a spinning space,
in the company of past experiences.
The wool is there,
open for the hands to listen to,
and they do.
They dig deep,
roll around,
delighting in the touch,
deepening
sensation and sensibility.
Frrp, frrp,
the yarn flips over the tip of the spindle,
connecting sound to the scene,
brain adding to the perception.
I put the spindle down
and close my eyes,
feel the vibrations of the fibers
like glitter
still dancing on my skin,
frrp frrps still swaying my eardrums.

I see the wool float
through the heart of the spinning
behind closed eyelids,
joy softly pulsing deep in my own heart.
Hands can't unfeel
what they have felt,
eyes can't unsee
what they have seen,
brain can't unknow
what it has made sense of.
Whenever wool washes over my hands
I become a new spinner.

In this chapter, I want you to learn to trust your hands to read the wool and your brain to store information about present and previous wools. The stored information is there for you to use when you explore future ones.

Hands and Brain; Sensation and Sensibility

The hands have an astonishing capacity to take in information. Through the ridges of the fingerprint and the winding of the ridges, they can register texture, softness, surface structure, and characteristics. Tactile perception is the superficial information we get through the hands' ability to register touch. When we investigate with our hands we move into haptic perception, where the hands' ability to register touch and movement allows them to survey deeper. The physical experience is registered by the hands and the sensory interpretations in the brain. When the hands feel a surface, the brain interprets it. Perhaps the brain remembers similar sensations or distinguishes the current one from others. The registrations from the hands are then stored in the brain. Just as the brain remembers structures and textures, it also remembers movement patterns. When we practice spinning with a specific spinning tool, the brain remembers and knows how to get back to that particular movement pattern. If the brain connects the sensation to a calming and positive emotion, oxytocin can be released and sink both heart rate and pulse.

The hands register size, temperature, moisture, dryness, brittleness, stickiness, oiliness, and viscosity. Physically touching things helps us explore and understand our surroundings (see also chapter 14). The cooperation of the physical registration of touch, the visual registration of shape and depth, and the brain's interpretation of sensory input can result in a complex combination of perceptions. The brain can even extend the sensation beyond the hands—when we use hand tools, the hands can "feel" the characteristics of the wool without being directly connected to the wool. If we add sound to the equation, the connections become even more complex. I have spent many hours in lecture halls and conference rooms with wool in my hands, spinning, knitting, or nalbinding; or spinning sessions listening to music

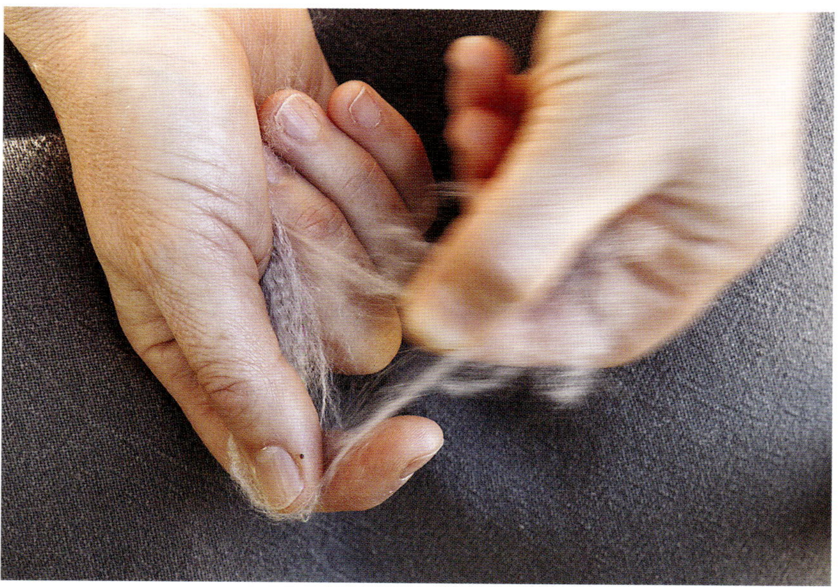

Figure 18.2. With the wool going through my hands, I learn about its characteristics and how it works in movement.

or audio books. If the brain stores the registration from hands and eyes, wouldn't that also mean that it connects the music, e-book, conference, or lecture to that wool in its bank of experiences? When I pick up that project the next day or the next week, my mind seems to take me back to the song, the speaker, or the narrator.

My fingers get to know each fiber in each new stage of the process—as collective members of a staple, as individuals in a teased blend of fibers and air, as a shaped structure in a rolag or top, and moving together in the draft to finally become locked as a community in the yarn. After that, the yarn works as a new unit to learn from in its behavior in a textile technique and a finished textile. For every new step, the brain recruits previously stored information to understand the new reality. The notion of the depth of information sizzling in the hands and brain at the processing of one single fleece is overwhelming. Still, this knowledge is available for us—over and over again—if we choose to work with our hands, if we choose to read the wool by touch. For every new fleece I dig my hands into, I add to my bank of experiences. For every new fleece I touch, I can dig into that bank and understand new dimensions that I couldn't before. I ground in my new understanding and explore from there.

When I approach a fleece, my hands are my most important tools. With the cooperation of the senses and their joint understanding of complex sensory information in mind, I believe in the power of being in the wool with my hands. I am convinced that the hands need to be in the wool and in the making for us to understand and remember.

A few generations ago, crafting was essential for self-sufficiency. We needed to know how to carve, sew, weave, and build for the household. Today, hands have been removed from the making of things and the creation of ideas on a larger scale through industrialization, digitalization, and consumerism. Perhaps we need crafting more for spiritual than self-sufficiency reasons—to sense natural materials and the joy of making in our hands.

Hands in Wool and Words

If I learn with my hands in the wool and understand it on a physical as well as a cognitive level, wouldn't writing the experience down enhance the experience even more? Over the years I have been spinning, it's like my brain has come to trust my hands to do the thinking for me when I am in the wool. Of course, it is the brain's interpretations of the signals from the hands that are working, but it is on a subconscious level. My hands relate to each other, to the wool, and to the tools. They plan the process but are always alert and ready to change the plans according to the whim and the whisper of the wool. They refer back to their experience while at the same time listening to what the wool in my hands is telling them, always open to rewrite the route. It is in this flow of making that I feel my brain can relax and follow the rhythm of the hands. This is where I can find the space between my thoughts.

What if, with the wool still vibrating in my hands after I have let go of it physically, I trust those same hands to write the wool, with pen on paper? The thought of it makes me giggle with amazement at my hands, those parts of me that are so wonderfully vital for my learning about and experiencing the world. When I write, I trust my hands to write whatever wants to be written and just go with the flow. Afterward, I hardly recognize what I have written. By putting the wool on paper through the power of my newly wooled hands, my hands trust the brain back to reflect on their findings and put them on paper in squiggly loops. If I squint, the loops may look just like the locks themselves.

Since the late nineteenth century, all Swedish children through compulsory education learn crafts within the subject *slöjd*. The word *slöjd* comes from an Old Swedish word that actually means "smart." The subject deals with crafting in wood, textiles, and metal. Every now and then debates arise about the importance of more theoretical subjects, and *slöjd* is on the chopping block in the curriculum to pave the way for what are considered more important matters. But what would happen if we took away the one subject where we learn through the astounding capacity of our hands? What would happen not only to the subject of *slöjd* itself but also to the understanding of the world around us, including those more theoretical subjects?

During the pandemic, I noticed a surge in the number of people around me who wanted to learn to work with their hands. Lots of people enrolled in my courses, I got inquiries to lecture online, yarn shops ran out of yarn, and the queue to an allotment in my community garden exploded with new members. In a situation where the world seemed to turn awry, people were eager to learn to provide for themselves and their families—growing a garden to feed the family, learning to spin or knit to keep warm. When the implications of industrialization, digitalization, and consumerism are put in a new perspective, we seem to want to go back to the hands that were taken away from the making of things and ideas. When we spin, I believe we keep the knowledge of the hands fresh and ready to serve us. *Slöjd* is smart. Let's keep the knowledge of the hands in the hands and *slöjd* on.

LISTEN: READ

Now I'd like you to read the wool with your hands. Pick any wool at any stage of the process and be in the wool and in the making. Read your wool. Tease with your hands, feel every strand as it leaves the staple in a bow and opens up into a new structure. Watch the transformation from staple to wool mass. Card your little heart out and feel the motions—the length and elasticity of the fibers between the tools—through your hands. Watch and feel the fibers separate before you. Investigate your hands' capacity to actually feel the wool through the tools. Sense the fibers glide past each other in the draft, the vibrations of the spinning, and the settling into the yarn. Be in the wool and allow your hands to read it and guide you through the process. Then, write the wool. Write what you just experienced, what you felt, saw, and heard. Don't bother with word order, sentence structure, semantics, or the grace of your longhand. Just write your little heart out and see what you get. Read what you have written. Do the whole exercise again with the first time fresh in your hands and your memory. See what you get this time. Thank your hands for serving you with their knowledge.

CHAPTER 19

With My Two Hands

In this chapter, I want you to learn to listen to your body and to the information you receive through the roles of the hands. This chapter has two parts, both of which emphasize the importance of using both hands in both hand roles and listening to what they tell you.

The first part is about spindle ergonomics—about always pulling the spindle toward the palm. This implies switching hands. The second part is about understanding both hand roles—the roles of fiber hand and spinning hand—with both hands and any spinning tool. The cooperation between the hands can have a function of a third task, one I like to call the third hand.

Pushing, Pulling, and Cramping

Around 2018, I was exploring the technique of in-hand spinning. The spindle is a shaft with a belly and usually a detachable whorl at the bottom. The spinner twirls the spindle in one hand and holds a dressed distaff in the other. Ring finger and middle fingers lightly clamp the spindle, while thumb and index finger roll the shaft. For short moments the spindle can spin freely by the momentum of the twirling, but most of the time it rolls in a one-to-one relationship with the twirling motions. The technique is often referred to as medieval spinning, but it is also still practiced in parts of France, Portugal, and the Balkans. As I am a leftie, it felt natural to spin with the spindle in my left hand, but since I wanted to use the yarn for knitting, I spun it clockwise, pushing the spindle out from my fingers. After ten minutes, I got a cramp at the base of my thumb. I kept trying and kept getting the same cramp. I asked in a Swedish spinning forum if someone could explain the pain to me. An occupational therapist replied and told me that we have about twice the number of muscles governing pulling movements compared to pushing movements with the hands. This is an evolutionary development—it is more beneficial for us to pull

Figure 19.1. Learning to use both hands as spinning hand and fiber hand can teach us a lot about the hand roles and about the cooperation between the hands.

things toward us than push them away. When you row a boat, you pull the oars through the heavy water and push them through the light air. With a repetitive movement pattern, more muscles will share the strain for pulling motions than for pushing motions. To spin my yarn clockwise with less strain I therefore needed to switch hands and use my wonky right hand as spinning hand, pulling the spindle into the

Figure 19.2. For a clockwise spun yarn, my right hand is the spinning hand.

Figure 19.3. For a counterclockwise spun yarn, my left hand is the spinning hand.

hand. After a shorter time than I had expected I got the hang of it, without pain. I realized that the concept would apply to other spindle types as well, and relearned with those too, clockwise with the right hand, counterclockwise with the left.

The notion of how the muscles work changed the way I spin, the way I think about spinning, and the way I teach spinning. I now speak fluent spinning with both my hands in both roles in every spindle technique I learn and practice. I make all my students learn to spin clockwise with their right hand and counterclockwise with their left hand. With supported and in-hand spindles, we practice flicking or twirling the shaft with the fingers pulling inward, and with floor spindles, we practice rolling the shaft with the flat hand moving up the outside of the thigh. For suspended spindles (with circular whorls), both finger flicking and thigh rolling work. In standing or walking spinning with a suspended spindle, thigh rolls work downward, too, on account of gravity. With a spindle where I can switch between an underhand or an overhand grip, such as some horizontal and diagonal spindles, I can use both hands as spinning hand if I change the grip: for a clockwise yarn, I can spin with an overhand grip with my right hand as spinning hand or with an underhand grip with my left hand as spinning hand. This is the case with the Azorean spindle (figure 20.1).

For another project, I spun from the cut ends of lightly teased staples (see chapter 7) on my spinning wheel. The preparation is a little less open than a carded or combed preparation and thus requires more work with the spinning (front) hand. I got a cramp in the thumb of my spinning hand and realized that I needed to either switch preparation methods or switch hands. I took the brave route and switched hands, moving back and forth between batches. It felt awkward at first, but after a while I got fluent enough in my new hand roles to be able to switch with reasonable comfort. I also learned a lot about what the hands do in their roles as spinning hand and fiber hand and about how the hands communicate through the wool.

Hand Roles

Elated by what I had learned from switching hands, I created a five-day challenge for my blog readers and students to encourage them to play with switching hand roles. In chapter 9, we established the following:

- The hand that holds the spindle or is in front in wheel spinning is referred to as the spinning hand. It manages the yarn and controls the twist moving into the fiber.
- The hand that holds the fiber is the fiber hand. It manages the fiber and controls the fiber moving into the twist.

You may also encounter such terms as the spindle hand for spindle spinning, and front and back hands for wheel spinning. Both hands have a light grip on their respective ends of the process, receive the fibers, and move them along in the flow from fiber supply to yarn.

The challenge required two spindles, one for each spinning direction, or a spinning wheel. For the challenge I invented new parameters: If my default hand positions are my regular hand roles, the roles for my switched hands will be the bonus hand roles. For wheel spinning, my regular spinning hand is my left hand and my regular fiber hand my right hand. Bonus hand roles for spinning on a wheel would thus be my right hand as spinning hand and left as fiber hand. This is what I practiced when I took the five-day challenge myself.

For spindle spinning, I can now honestly say I don't have regular hand roles. The direction I'm spinning in determines my hand roles.

Hand Roles with the Great Wheel

For a great wheel (walking wheel), the hands are fixed and the roles slightly different. The wheel is constructed for the left hand to hold the fiber and the right to turn the wheel (figures 14.2 and 20.2). This would technically mean that the left hand is the fiber hand, with the spindle as its resistance. There is no spinning hand. The fiber hand can take over some of the spinning hand roles, though—in the way it controls the yarn angle from the spindle, the amount of wool that goes into the twist, and the twist angle. There is a progression for the fiber hand to take over parts of the spinning hand's role, from an English long draw (see chapters 9 and 10) on a spinning wheel, through the floor spindle, and to the great wheel:

- With the English long draw on a spinning wheel, my spinning hand usually opens up the twist (see chapter 9) repeatedly and lightly for a smooth flow.
- On the floor spindle (or other spindles worked diagonally from a surface), I may place my spinning hand thumb on or under the yarn to open up the twist, but it is not constantly on the yarn. The yarn angle (controlled by the fiber hand) plays a role in the yarn quality.
- With the great wheel, there is no spinning hand on the yarn (unless I stop the wheel and troubleshoot).

Switching Hands

"But I'm so right-handed, I can't possibly spin with my left hand," you may say. And I say, "I don't buy it," in general at least. If you have ever held fiber in your left hand, you can spin with it, well, too, if you practice. The thing is, we use both our hands when we spin, and the roles of both hands require fine motor movements. So, when you spin (clockwise) with your right hand as spinning hand, you handle the wool with your left hand. You may just not think about what handling the fiber means. You will soon.

Many people seem to focus on the spinning hand and talk about spinning as "I spin with my *x* hand." However, we spin with both hands. The fibers flow from the fiber hand, through the spinning triangle, and through the spinning hand before they are rolled onto the bobbin or shaft. Meanwhile, the twist flows in the opposite direction, from the spinning hand through the spinning triangle toward the fiber hand. The movements meet at the heart of the spinning. In the draft, the hands cooperate to steer twist and fibers where they need to go for the quality you are working to achieve. This is what we talk about in the Twist Model in chapter 9. The heart of the spinning may seem like a romantic term, but as a point that pulses between the hands and between fiber and twist, I find it a useful metaphor. Both hands have equal part in and responsibility for keeping the spinning heart beating smoothly.

For the fiber hand to be able to move the fibers from the preparation to the spinning triangle, it needs to be in control of the portion and distribution of fibers moved forward, for every draft. This essentially means that the fiber hand is responsible for the thickness and the evenness of the yarn. If I encounter an issue with the yarn, I always take a step back to find where I have missed something. And the first step back in the fiber flow would be from the spinning hand to the fiber hand. Perhaps I am holding on too tightly to the fiber, making it felt in my hand. Perhaps I have served too big a portion of fiber into the spinning triangle, creating a lump in the yarn. The action can go in the flow of the twist, too. If the spinning hand allows too much twist into the spinning triangle, it will be difficult to open up the twist (for woolen spinning) or to keep the twist out of the drafted fibers (worsted spinning). Usually, I need to stop and untwist or increase the distance between my hands until the fibers can move comfortably again.

The first exercise in the five-day challenge I created was for the students to observe the tasks of their regular hand roles. Many students were surprised at how much the fiber hand actually does. It seems like we rarely talk about it. As the students moved on to switching hand roles, many of them expressed a frustrated feeling of being a beginner at a craft they know. Bonus fiber hands awkwardly handling the fiber, bonus spinning hands allowing too much twist into the fiber, all the while they know what is supposed to happen. Hands and brain feel out of synch. The students got to try to narrate their spinning (see chapter 15) with their bonus hand roles. Bit by bit they seemed to get their hands to communicate with each other and cooperate for the sake of the wool between them. In chapter 18, we talked about how the brain can remember movement patterns. This happens here, too, and the more we practice, the better the bonus hand roles can learn their new tasks. With the knowledge from the regular hand roles of what is supposed to happen, we get to look at ourselves from an outside perspective. Meanwhile, we experience it from the inside. An opportunity to learn, no doubt!

Why Switch?

There are many benefits of switching hands and of learning to spin with both hands in both hand roles. As exemplified with the hand-switching challenge, we can make ourselves aware of what the hands do. We can also learn to work more sustainably—to present an even fiber portion from the fiber hand and even twist from the spinning hand. The awareness of an even workload for both hands in both roles can result in a more consistent yarn in either choice of hand roles.

If you are a knitter, a counterclockwise spun and clockwise plied yarn is usually preferred, and you naturally would spindle spin this with your right hand as spinning hand and ply it counterclockwise with your left hand as spinning hand. If you want to knit with a singles yarn, you may want to spin it counterclockwise with your left hand as spinning hand, just as you would for a Z-plied yarn for two-end (or twined) knitting. Weavers may want warp and weft yarns (or sections of either) in different directions for reason of design or weave density.

In 2021, I designed a shawl for handspun singles. Knitting with singles can be a challenge, since there is a risk of bias in the fabric due to the energy in the unbalanced yarn. I decided on stripes in the shawl, with one color spun clockwise and the other counterclockwise, and a balanced knitting technique. I chose my favorite tool for bulky singles: my floor spindle. The experiment resulted in a shawl pattern, "Cecilia's Bosom Friend," published in *Spin-Off* magazine (see appendix 1).

I teach all my students in my in-person spindle classes to practice with both hands in both hand roles, regardless of their previous spinning skills. Experienced and beginner alike learn both ways. Even if they mutter while they learn, they do learn. As a spinning teacher, I find it valuable to have relearned switching hand roles—when I see

Figure 19.4. I have spun the white and gray singles yarns in different directions to keep an even workload on my hands and the knitted fabric balanced.

my students' wobbly attempts to learn the bonus hand roles, I can honestly say I know the feeling. By relearning, I will understand better how to teach.

The Third Hand

When I teach, I talk a lot about spinning ergonomics and what we learn by switching hand roles. One thing we spend a lot of time exploring is the cooperation between the hands. I like to call this process "the third hand."

For the fibers and twist to meet smoothly at this point, the hands need to cooperate in the brain. The fiber hand needs to adjust the fiber feed to the yarn that comes out on the yarn end. The spinning hand needs to control the twist that comes into the fiber. Both hands need to cooperate for the draft, with or without twist in the fibers. When they succeed, we can find a space where the spinning is easeful and gentle on hands and fibers. As soon as I need to open up the twist with more than a simple roll of the spinning hand thumb, there is too much twist or too much fiber in the yarn. When this happens, we either pinch and strain or stop and adjust.

The heart of the spinning is the physical connection between the hands that also carries the information between them, about each other and about the wool. When the fibers can move in the semi-stable part between the hands, they are free enough to transmit the signal between the hands (or, with the fibers drafted with no twist between the hands). At this point, where twist and fiber meet, we receive the benefits of preparing the fibers with those three grains of sanding paper we talked about in chapters 4, 7, and 8 (or pay the price of insufficient preparation).

To find the signal from the fibers between the hands, they need to keep the fibers under adequate tension. Through the tension, one hand can literally feel what the other is doing. When we add our gaze to the equation, we get even more information to the brain about what is happening and how we can plan for the smoothest ride of fibers from preparation to yarn. In a spinning technique where the hands are far away from each other, like the floor spindle, the hands depend on the tension between them to carry the information about the arm's-length semi-stable part, and even more so with the great wheel where one end of the tension is the spindle.

The final exercise for the students in the five-day challenge was to switch back and forth between regular and bonus hand roles. Many of the students were fascinated by the feeling of their hands "talking" to each other between the switches—the regular hand roles guiding the bonus hand roles into coordination and control while the bonus hands could teach the regular hands to relax more. Together they could work for a more consistent yarn and easeful spinning. Switching hands can be a gift to yourself, ergonomically, functionally, and for the cooperation, the third hand.

LISTEN: SWITCH

For this exercise, I would like you to play with switching hands, working ergonomically and learning about the hand roles and how they can work together for ease and quality. This exercise can take time. Do give yourself that time. Practice for fifteen minutes a day, take notes, and let the brain process the new experiences overnight.

If you spin with spindles, use two: one for each direction. Always spin clockwise with your right hand as your spinning hand and counterclockwise with your left hand as spinning hand. If you work with a spinning wheel, switch between your spinning (front) and fiber (back) hand.

Start by using your regular hand roles. Observe what they do and pay extra attention to the fiber hand. During the next few sessions, switch to your bonus hand roles. Just observe at first, letting the hands understand the new situation. Move on to reflecting more deeply about what the hands really do. Go into details here. Look at the notes from previous sessions. For the next step, I encourage you to narrate your bonus hands' spinning (see chapter 15). Talk about what you do, what you see, and what you need to do in relation to what you see. In the final step, switch back and forth between the bonus hand roles and regular hand roles. Use an interval that suits you: time, fiber unit, your favorite song. Reflect over how the hands cooperate—fiber hand and spinning hand, but also regular and bonus hands. Focus on the wool that links them together. What can the hands learn from each other? Do you need to backtrack to the fiber preparation? Take notes and compare these to your earlier notes from this exercise.

This may feel awkward. Your brain knows what is supposed to happen, but your hands are just floppy fish. Keep practicing, be kind to yourself, and ease your expectations. Eventually you will see progress. We can learn a lot about the comfort of spinning by visiting the discomfort. You can do this exercise several times, for different spinning tools and for different kinds of drafts. Allow it to become part of and natural in your regular spinning practice.

CHAPTER 20

The Inner Spinner

While I may be curious about an idea or a technique for a fleece, the inner process is what I keep coming back to—the feeling in body, mind, and spirit as I spin. In this chapter, I invite you to find your inner spinner and the different processes that contribute to your experience of spinning beyond making yarn.

Rhythm

The spinning process is for me a whole spectrum of processes in different dimensions. One way for me to find a lightness in spinning is in the rhythm. If I pay close attention, I can find different kinds. Perhaps I find a rhythm in the treadling of the wheel or in the flicking of the spindle. When I card, I choose a number of strokes in each pass, and a number of passes. This keeps my carding consistent, but it also gives me that extra dimension of rhythm, something for me to keep in my mind's eye. With a three-stroke, three-pass carding rhythm, I do the following:

> Dress-two-three-softly-stroke-six-even out
> card-two-three, transfer-twice,
> card-two-three, transfer-twice,
> card-two-three and transfer-twice.
> Lift-tuck-lift-tuck-lift-tuck,
> lift and go back, roll-it-home.

Figure 20.1. There is a sweet choreography to spindle spinning. How do you dance your spindle?

There can also be a rhythm in how the fibers go between the hands in the heart of the spinning. The ongoing pulse of twist and fibers meeting is a rhythm in itself. There can be a rhythm in the way I prepare a bobbin-size portion, spin it, and repeat for an entire fleece, or in changing the order and preparing, spinning, and plying for one skein and going straight to knitting it (see chapter 12). The overarching rhythm of practicing—making mistakes, learning from them, and moving forward—is also a rhythm worth mentioning.

Most of my spinning rhythm is in common time (meaning four beats per measure in musical terms); it always has been. If I walk while I spin on a suspended spindle, there is a rhythm in my steps and in my breathing. If I spin an English long draw (chapter 9) on my spinning wheel, I find one treadle count for gathering twist and one for adding twist. Usually, I add or subtract to land in chunks of four. Once, I found myself treadling in triple time and, as trivial as it may seem, it was a totally new experience. There is a lightness in triple time that I hadn't experienced in common time. I found myself dancing the spinning waltz and embraced every sway of it. An added layer in the rhythm surfaced. I let my expectations of the end product go and rejoiced in the making. I experimented with actually listening to a waltz while I spun and dived even deeper into the rhythm of the spinning, working with gentle hands and light movements.

Our brains coordinate the input of multiple senses by connecting them in specific areas of association and processing these together with previous experiences. With this multisensory cooperation, we can deepen the sensation of the making. Musicians work with a continual connection between sensory and auditory input from hands and ears. Spinning is not that far from making music. Just as the musician molds music out of the instruments with their hands, spinners shape the yarn with the same fine motor movements. If we then connect rhythm and music to the spinning process, we might have a deeper experience, especially if we have practiced enough for our brains to have created motor programs that remember and steer the movements of the hands. Going back to the discussion of learning to spin with both hands in both hand roles, I envision the brain creating even more motor pathways to strengthen the spinning web. Imagine that—a whole woven structure in the brain created by the creating of yarn. When I read research about musicians' higher developed white matter—the communication coordinator of the brain—I can't help thinking it may be relevant for spinners, too.

Dance

I attach the end of the rolag to the yarn on the spindle and make a join. I turn the wheel at my side as I watch the twist enter the wool, *arrière* in adagio, one, two, and three steps back. I angle the yarn, arm elongated and open to the world, with the wool stretching like a rubber band. Twist moving up the wool, moving up the wool, sizzling from the tension between the spindle and my left hand, through my chest to my right hand, dancing the wheel. I feel the twist entering the fibers, securing them into a gentle hold. With another few turns of the wheel, its spokes paint a shadow dance across the newly mowed lawn and my bare feet. With a soft hand, I change the angle of the yarn again, take one step right and turn the wheel. I take two steps *en face* to roll the newly pirouetted yarn, twirling in its own dance, onto the spindle and settle the yarn onto its spool-shaped nest.

As I take the next set of steps back and make that yawning draft, I watch another dance, in the wool right before me—not still open fiber, not yet yarn, but in that magical midway, vibrating with possibilities. I hold the section in its in-betweenness, the outstretched heart of the spinning, until I'm happy with shape and twist. On and on we dance—wool, wheel, hands, and eyes, feet treading lighter as my body learns the choreography. I keep walking three steps back, spinning the wheel, one step right and two forward, bare feet rolling on dew drops, mindfully from heel to toes. The lawn is my dance floor. I invite the oaks around it to twirl with me, wiggle their roots, branches reaching to the sky. The transformation of the wool, the dance in my hands, and the steps of my feet expand my vision and alert my senses. Motions light, grip gentle, and pace steady, through rhythm, twirl, and dance, I enter my inner spinning ballroom.

All the steps from wool to yarn involve some sort of movement, and any movement can be turned into a dance. Spinning and wool preparation can have aspects of both a set choreography and free movements. I may have a sequence in the way I work, but I also need to be prepared to improvise, pay attention to the cooperation between my hands, and switch between leading and following the wool.

Figure 20.2. In walking-wheel spinning, the process asks the whole body to dance.

Interpreting spinning through dance deepens the experience for me and eases any expectation of results. The dance is here and now, and for me to explore and enjoy. How do you dance your wool, I wonder?

Memory

I place my bare feet on the treadles and let my shoulders sink. Meanwhile, my mind picks up threads of memories, of sensations and images from earlier spinning moments. The words and voices of Barbara Kingsolver, Jonas Hassen Khemiri, Robin Wall Kimmerer, and other writers whose audiobooks I have listened to through the creation of this yarn, quietly announce the next chapter in my mind. It's funny, as I make that first turn of the wheel to start my spinning session, I am thrown back into the book again, the words nestling their way into the yarn, yarn into the words, and staying there. Later, as I wear the sweater or lean against the pillow I made with this yarn, I am wrapped once more in the words, voices, stories, and landscapes that accompanied me through creating.

We have talked about listening to music while spinning and experiencing the rhythm, but the memory of music can also be spun into the yarn, just as the audiobook. I may spin on a train, while attending a lecture, at a coffee break at work, or while just allowing my thoughts to come and go. The memories of whatever goes on around me as I spin are collected in the spinning; I literally spin them into the yarn. When I stop, they linger in body and mind. Whatever state of mind I have been in while spinning is gently stored in my heart in a shield of wool. Good things, bad, happy, and sad—they are all there and part of me. The finished textile becomes more than itself, a vessel for all the memories stored in it.

Creativity

The intelligence of the brain and the hands are linked to each other, and the hands have an important role in learning. The movements and activities of the hands can also spark inspiration. Oh, all the writing I have done while spinning! When I am in the flow, with my fingertips automatically scanning every aspect of the wool, I often find solutions or insights to challenges I have been contemplating. It can be a deeper thought process regarding the spinning itself or something else. Rushing to my notebook for emergency writing, I jot it down and make the reflection clearer to myself. With the thought on paper, I can go back to the spinning and test my theory on a deeper level. I have created many blog posts this way, learning from my hands in the wool, writing, realizing, going back to test, then writing some more. The satisfaction of seeing a fuzzy reflection more clearly can be profound. Sometimes I get attacked by new projects when I spin, sparks that knock on my brain and want to be molded into a mature idea and realized.

I can also work this the other way around. If I find myself blocked in my writing, thinking, or planning, I can turn to whatever spinning project I have going (usually more than one), spin for a while, and feel lighter as the block dissolves. Spinning opens doors to other creative endeavors, and I often switch between spinning and writing to open creative blocks.

LISTEN: DANCE

Today I invite you to dance your wool. If dance doesn't appeal to you, choose another way of being in the creative process. First, find a spinning space where you feel safe and at ease (find inspiration in chapter 13). Perhaps you have one already; perhaps you will create one. Make it inviting, enjoyable, and beautiful.

Sit in your dedicated spinning space. Close your eyes if you want to and take a few breaths to center yourself. Start spinning (or preparing wool). Try to be in the process, working in a flow without judging. Make observations of what happens and what you experience; be curious and open to right now. If you find something challenging, stay in it. Explore it and reflect on why you find it difficult. Give yourself time to be in the wool and in your hands. Perhaps you feel a rhythm or a dance. Go with it; let it sweep you off your feet. Perhaps your mind takes you to the last time you worked with this wool or this tool. Explore the memories. Perhaps you find a solution to something you have struggled with; perhaps you get an idea. Explore it. You may just sit with your process, letting thoughts come and go. If you want to, find a space outdoors and enjoy the new dimension of spinning with wind in your hair and the sky as your roof. Be in the process for at least fifteen minutes if you can.

End your session by closing your eyes and taking a few breaths. Make a few notes. Thank yourself for taking the time to be in the spinning and exploring it without expectations.

If you want to, you can use this exercise to create a spinning ritual for yourself that you can come back to again and again. Perhaps start or end every day with a few minutes of quiet spinning.

CHAPTER 21

Let the Wool Be Your Teacher

It starts in the living room of my mistress, the pasture where she grazes, lives, sleeps, and nibbles grass and herbs, keeping the landscape open. She fertilizes the soil and keeps it alive and well. Insects thrive on the plants that are typical for this pasture, the plants that depend on her. By just being in the world, she and her flock sisters manage the landscape and help preserve biodiversity.

Go Back to Move Forward

Through millennia, sheep have grown wool developed to suit the specific environment they are in. Sheep are the source of so many good things in the world: manure in an allotment, cooler bags with wool as insulation, lanolin to soothe sensitive skin, sheep's milk ice cream on a sunny day . . . and wool gliding through hands, warming hearts yesterday, today, and tomorrow. I am the wool, a product of a living being and endless generations of wool evolution.

Every year I grow out and renew, in service of my mistress. Perhaps you have even met her, buried your hands in her soft neck curls, and smelled her breath next to yours. When she is finished with this season's edition, you are welcome to that, too. If you lean in, you can read the events throughout her life—sunny days with fresh grass in her bellies, needles from the juniper patch in the southern slope of the pasture, cold January mornings with tiny hooves kicking her from within. It is all in my fleece to learn and read, through the length of each of my staples.

Take Your Time

Pick one of my staples; my treat. Greet it in its own shape. Right there, the second you separate it from its staple sisters, I give you the first clues to what my fibers are about. Perhaps my left flank staples are a little tousled; it may be my mistress's favorite napping side. Perhaps a seed or a piece of moss falls out and invites its equivalent of air. Perhaps your hands are softened by the lanolin. Again, my treat. Embrace me in all I have to offer. To find out more, keep exploring with open eyes and open heart. You see, I don't speak your language, but you can learn some of mine.

Let's move on. What do you see in the staple? What do my fibers look like? Can you see long and billowing fibers, curiously stretching their way beyond, clustering tight at the tips? Are they strong and shiny, allowing spring rains to slide along their length, away from my roots, shielding the sheep from the downpour? Can you see soft and airy fibers, warm, fine, and expanded? Or perhaps you see both in the same staple, forming the perfect shield against the elements. My mistress made that. Can you feel how my fibers work in movement? Do they glide smoothly when you pick the staples, or is there a resistance? Is the glide swift like a summer breeze or viscous like honey? I can do both! What happens when you invite air between my fibers? Do they behave differently? Is the glide lighter, the resistance lower? Is there more trace of the pasture in your lap?

Perhaps you find favorite parts of me—the airiness of my undercoat, the shine I use to dazzle the world, or the crimp of my curls. I have many qualities to offer. With your hands in my fibers, you can translate my assignment to grow a protection for my mistress into your wish to spin a yarn to protect yourself or a loved one.

In the book Momo by Michael Ende, the girl Momo can find answers to people's problems. She does this by just being with them and listening to them. The Men in Grey, eventually revealed as a species of paranormal parasites stealing the time of humans, spoil this pleasant atmosphere, though. One of the most important steps Momo takes in winning the stolen time back is to walk backwards. Only then can she move forward. To come to the end

Figure 21.1. Trust your hands to listen and let the wool be your guide.

of the yarn you want to spin me into, go back to the beginning, to my raw fleece. Get to know me, trust me, and let me lead the way forward.

Look at all the gifts you have received already: the gift of my wool from the sheep, of learning from my characteristics, and of learning the craft, of the mistakes you make, and of the process in your mind and your body. You can reciprocate these gifts by taking your time to listen to what I can teach you, and spin me into beauty.

I have one more gift to you, and it is right below. Prepare some wool and choose a spinning tool before you start. ✻

Listen to the Wool

Sit down, make yourself comfortable. We are going on a spinning journey, you and me. Take a deep breath. Exhale through your mouth; relax your shoulders.

Start spinning. Get acquainted with my wool and the spinning tool. By now I may have flowed through your hands many times. Spin me gently; listen to my fibers. They will guide you if you let them.

Feel the fibers in your fiber hand. Feel your hand feeding the fibers toward the spinning hand. Is there a difference in how you hold the fibers at the outer and inner ends of your hand, where the fibers are open and where they move toward the draft? Can you feel the thickness of each portion you serve to the spinning hand? Are they the same or different? It is okay if they are; just observe. Let your fiber hand get acquainted with the flow of fibers you move forward. Trust your hand to decide what to do.

What do you feel in your spinning hand? How does it receive the fibers? Are they coming in an even flow? Is the flow smooth, or do you need to steer the fibers? Stay for a while; observe what you do with your spinning hand.

Feel the twist coming from the spinning tool. How does the twist meet my fibers? Do you allow the twist into drafted or undrafted fibers? How do you sense my fibers responding to the twist? Do they fall smoothly into it, or do they seem to have their own will?

Dance my fibers through your hands; let the hills and valleys of your fingerprints be part of the dance. Let them explore my length, crimp, glide, resistance, and fineness. Can you sense the flavors of my fibers in your hands and between them?

The heart of the spinning, that short section between your hands, is what makes the connection between them possible. The sensation of the fibers in your fiber hand is telling the spinning hand what is to come, through my connection between the hands. The spinning hand in turn feeds information back to the fiber hand, perhaps asking it to slow down or speed up. I can feel the conversation of your hands as I stretch between them. Attentive voices: whispering, serving.

The information is here if you trust me to hand it to you, if you trust your hands to listen.

Ask me with your hands,
ask me how long my fibers are, and I will show you.
Ask me how my fibers glide, and I will guide you.
Ask me how my fibers move,
and I will take your hand and lead the way.
You just need to know how to ask, how to
find my voice.
Let it sing with yours in the heart of the spinning.
You see, this is where it all comes together–
fiber and yarn,
twist and relaxation,
fiber hand and spinning hand,
and the way I dance through it all.
This is the space where my heart is close to yours,
beating in the same pace.
Back and forth,
wool to yarn,
flowing,
surrendering to the twist,
steered between your intelligent hands.
In that joint heartbeat
we find a united flow of fibers,
of thoughts coming and going,
without judgment.
We spin together you and I,
leading,
following,
listening,
in mutual understanding
and reciprocity.
Spin me gently,
spin me wildly,
let me trickle or gush,
joyfully through your fingers
in any pace,
any expression you like.
Let my fibers be breathed between your hands,

Figure 21.2. Go back to move forward: back to the sheep and the characteristics of the wool as a shield against the elements. Listen to the wool and let it be your teacher.

flowing in their own spirit.
By exploring,
playing,
creating,
expanding
and listening,
you have created your own bank of knowledge,
deeply woven
into my most beautiful yarn.

This moment, this spinning and all that went on around you, are forever spun into this yarn—happy memories and sad, they all deserve a place in it. They were part of making it, part of your heart and soul as my fibers flowed through your hands and mind. My soul and yours are intertwined in this yarn. There is a secret bond between us now.

Spin in stillness for as long as you like. If you want to, close your eyes, or let them be soft. Trust your hands to listen, to feel their way into my heartbeat. This is your moment. Make it joyful.

When you feel ready, slow your spinning down. Breathe. Feel. Can you still sense my fibers moving through your hands, even though our spinning has stopped? Can you feel the process lingering in your mind? Stay in this space for a while; let the movement settle before you take in your surroundings again. Take another breath and slowly open your eyes.

You are a new spinner now. ⁜

LISTEN: TRUST

This chapter is a spinning meditation. Trust the wool to guide you; trust your hands to listen to the wool. I recommend doing the meditation with a spindle.

You can listen to a twelve-minute audio version of the meditation (in English and Swedish) via this link: https://waltin.com/spinningmeditation/

Please do not share it.

EPILOGUE: ECHOES

As I lay down my spindle, the spinning is still in my hands. If I close my eyes, I can feel the structure of the wool as it becomes yarn right between my fingers. Opening my eyes, I am surprised to see my hands empty. My heart sings as it remembers all the moments I have spent with this wool. All I have learned; all the mistakes I have made. All the thoughts that have come and gone; a breath moving in and out, the wool passing through my mind on its way to a new shape. Like an echo, softly whispering its secret message, if I lean in and listen.

Back in 2016, when I first learned to spin on a supported spindle, I wanted to explore more supported spindles. I wrote in an online forum and asked for guidance on what I should look for in a supported spindle. I got some replies, but they weren't really what I had been hoping for. Someone suggested I go to a fiber festival to try out supported spindles from different makers. A wonderful idea, of course, but there were no fiber festivals in Sweden at the time, and no makers of supported spindles. I let the idea rest for a while. A few days after the discussion had

Figure BM.1. Receive my gift and pay it forward.

ebbed out, I got a private message. Sandy in Georgia wrote that she was willing to "send me some spindles." I paused. Why would someone want to send spindles to a stranger across the pond? "I have too many, and it's a bit embarrassing. See it as a random act of kindness and pay it forward." Her reply swirled in my mind as I waited for the parcel to arrive, not entirely sure it would. But it did, stamped with colorful butterflies and cushioned with a braid of wool to match their fluttering wings. Tucked between the fibers were eight supported spindles. My jaw dropped. I was standing among a range of supported spindles from different makers, in my own at-home fiber festival.

I needed to find a way to pay this astonishing act of kindness forward. I decided to stretch the at-home fiber festival a bit further. I created what I called a "traveling spindle library." I kept the fiber and two of the spindles, and added a couple of my own spindles and some fiber to the library. My idea with the traveling spindle library was to send the spindles to someone in my Swedish online spinning community, to someone I didn't know but who had shown interest in supported-spindle spinning (which was quite new in Sweden at the time). The receiver would then try the spindles for as long as they wished, keep one or more if they wanted, add some back to the library if they wanted, and then send the library to its next librarian, anyone they felt would benefit from a traveling spindle library.

I contacted the spinner I had chosen to be the new librarian, Elaine, and asked if I could send her the library. She showed the same kind of skepticism I had when Sandy in Georgia had contacted me. I explained the idea as straightforwardly as I could. Elaine was concerned about the money, but I assured her that it wouldn't cost her a thing. At last, she gave me her address, and I sent the traveling spindle library to its new librarian. This was back in 2016, and the library traveled from spinner to spinner in Sweden for quite a few years.

I have been teaching supported-spindle spinning for many years now, and in every class, I tell the story of the spindle library. In every class, I am truly grateful for what I have learned from teaching, from the questions my students ask, and from getting to know their individual learning styles. Writing this book, sharing my words as beautifully as I can, wrapped in imaginary colorful butterfly stamps, is my way of paying these gifts forward.

Keep exploring the wool you have, play with techniques in ways you hadn't planned, create from your heart, and expand your spinning horizons from the tools and skills you have. Keep listening to the wool, allowing it to be your teacher. And if you enjoy the gifts of this book, try to find ways to pay them forward. Perhaps teach a friend to spin, make something of your handspun yarn and give it to someone you love, or share your own words. Let what you have learned create an echo that reciprocates the gifts you have received. Make it joyful.

Happy spinning!

ACKNOWLEDGMENTS

When I created the outline of this book, I added an acknowledgments section and pondered about what to write there. All the acknowledgment sections I have read look so grown-up. At that point, with only a dream of a book, I could hardly envision the day I would have a finished book in my hands. As I have been writing, I have added names to the section, and it has become longer chapter by chapter. It turns out that I have many people and sheep to thank for the creation of this book:

Beth Kempton for numerous writing courses, gentle guiding, and the flowing of words. Isabel Atherton, my agent, without whom there would have been no book deal and, frankly, no clue. Candi Derr at Stackpole for believing in my idea and for midwifing me as a first-time book mother all the way to my book baby. Also Alden Perkins, Nancy Syrett, and others behind the scenes at Stackpole who in any way have been part of this journey. Jessica Linde, Sara J. Wolf, Barbro Heikinmatti, Anna McNaughton Lindemark, and Wendy Clark for endless support and for patiently reading through and strengthening my book proposal. For invaluable fact checking, grammar radar, and fairy godmothering: Deborah Robson. For guidance and fact checking about Swedish wool and sheep breeds: Alan Waller. For believing in me as a writer and for giving me various article and patternmaking assignments for *Spin-Off* magazine through the years: Kate Larson.

Claudia Dillmann, Margau Wohlfart-Leijdström, Fia Söderberg, Lena Hansjons, Milis Ivarsson, and Louise Westerberg and partners for generously opening the gates to your pastures and for sharing stories of your lives with your flocks. Arne Albertsson for hosting us in your log cabin and for the blissful silence of Tiveden, where many words flowed and where many of the photos in this book were taken.

Nilda Callañaupa Alvarez for kindly answering my questions about spindle spinning. Björn Peck for making the most exquisite spindles for my courses. There would be no teaching without them. Kia Gabrielsson Beer for holding my hand as I became a spinner all those years ago. Anna Herting, Boel Dittmer, Kristin Jelsa, and Ellinor Landmark, my wool traveling club friends for numerous wool journeys, laughs, and wooly hearts. All my students and readers, in-person and online. This book wouldn't have been in your hands without your support, your questions, and the reflections we have made together in the classroom. Really.

Cecilia, my sister in wool for endless conversations and deep friendship, and to whom I dedicate this book. Dan, my love, for asking the difficult questions; for your sense of light, harmony, and composition; for acknowledging and encouraging my need to create; and for making my heart sing. Isak and Nora for brightening my world and for tolerating living in a house of wool. My parents for bringing me up in a creative home with an endless supply of crafting materials and imagination, and for always cheering me on.

Finally, I thank all the sheep who have welcomed me into their living room, shared their wisdom, and offered me their coat.

APPENDIX 1: FEATURED TOOLS AND WOOLS

Spindles

Azorean spindle (figure 20.1, page 147)
Björn Peck (figure 16.1, page 125; figure 16.2, page 127; figures 19.1–19.3, pages 141–142)
Bosworth (front cover and figure 15.5, page 123; figure III.1, page 82; figures 11.3–11.10, pages 100–101; figure 12.1, page 104; figure 12.2, page 105; figure 14.1, page 114)
Jenkins (figure 18.1, page 136)
Pushka (figure 15.1, page 118)
In the spindle rack on figure 15.3 (page 121): Jenkins (2), Kokovoko, Anne Grout, Andean pushka bought from Abby Franquemont, Louët, Traditional Armenian spindle gifted from Irene Waggener, Bosworth (3), Kundert, Wildcraft (2), Andean chac-chac bought from Abby Franquemont, Azorean spindle bought from Rosa Pomar, Magpie.

Spinning Wheels

Antique walking wheel, maker unknown (figure 14.2, page 115; figure 20.2, page 149)
Kromski Mazurka, Pre-production (figure 10.1, page 91; figure 13.1, page 108)
Kromski Symphony (figure 11.2, page 99)
Majacraft Little Gem (figure 10.2, page 92)

Preparation Tools

Gammeldags midi maxi combing station (figures 8.12–8.16, pages 75–77; figures 8.19–8.21, pages 78–80)
Gammeldags mini combs (figures 7.5–7.8, page 64; figure 13.2, page 110)
Louët flicker (figure 7.4, page 63)
Used cards from the 1980s. Leather cloth, medium tpi, maker unknown (figure 7.3, page 63; figures 8.1–8.5, pages 68–71; figures 8.7–8.10, pages 72–73)
Villa Laurila 108 tpi cards with leather cloth (figure 8.6, page 71)

Breeds

Åsen wool (front cover and figure 15.5, page 123; figure 2.11, page 21; figure 2.14, page 24; figure 5.3, page 51; figure 6.2, page 54)
Åsen/Härjedal wool (figure 6.3, page 56; hat in figure 21.2, page 154)
Brännö wool (figure 1.2, page 6; figure 2.1, page 12; figure 2.12, page 22; figure 2.18, page 28; figure 7.1, page 61; figures 7.5–7.9, pages 64–65; figure 8.1, page 68; figures 8.3–8.5, pages 70–71; figures 8.7–8.11, pages 72–73; figure 12.2, page 65; figure 14.3, page 116; figure 17.3, page 132; figure 19.3, page 142; figure 20.1, page 147)
Dalapäls wool (figure 2.10, page 20; figure 4.4, page 40; figure 5.1, page 47; figure 7.10, page 66; figure 12.1, page 104; figures 12.4 and 12.5, page 107; figure 14.2, page 115; figure 18.2, page 138; figure 20.2, page 149; figure 21.2, page 154)
Finull wool (figure I.1 [the sweater I'm wearing], page 1; figure 2.7, page 17; figure 17.1, page 130)
Finull/Rya cross (figure 2.3, page 14; figure 6.4, page 58; figure 6.5, page 60; figure 16.1, page 125; figure BM.2, page 173 [sweater])
Fjällnäs wool (figure III.2, page 84)
Gestrike wool (figure 2.2, page 13; figure 2.9, page 19; figure 2.15, page 25; figure 2.19, page 29; figure 2.20, page 32; figure 3.1, page 33; figure 3.2, page 36; figure 4.2, page 34; figure 4.7, page 43; figure 8.6, page 71; figures 9.2–9.4, page 86; figures 15.2 and 15.4, pages 120 and 122; figure 19.1, page 141; figure 19.2, page 142; figure BM.1, page 156; figure BM.2, page 173)
Gotland wool (figure 1.1, page 4; figure 1.3, page 7; figure 2.5, page 15)
Gute wool (figure 1.1, page 4; figure 1.2, page 6; figure 1.3, page 7; figure 2.4, page 15; figure 2.17, page 26; figure 5.2, page 50; figure 10.5, page 95)
Helsinge wool (figure II.1, page 46)
Jämtland wool (figure 1.1, page 4; figure 2.13, page 23; figure 10.2, page 92)

Klövsjö wool (figure 2.16, page 26; figure 6.3, page 56)
Roslag wool (figure 4.7, page 43; figure 9.6, page 88)
Rya wool (figure 2.6, page 16; figure 6.1, page 53; figure 11.2, page 99; figure 12.3, page 106)
Svärdsjö wool (figures 4.1 and 4.3, pages 39 and 40; figure 4.6, page 43)
Swedish Poll Merino (figure 1.3, page 7; figure 2.15, page 25; figure 9.1, page 85; figures 10.3 and 10.4, page 94; figure 13.1, page 108)
Swedish Leicester wool (figures 8.12–8.18, pages 75–78; figure 8.21, page 80; figure 13.2, page 110; figure 17.2, page 131)
Tabacktorp wool (figure FM.1, page iv)
Värmland wool (figure 2.8, page 18; figure 4.4 [mittens], page 40; figure 4.5, page 41; figure 4.7, page 43; figure 5.1 [mittens], page 47; figure 6.4, page 58; figure 7.3, page 63; figure 7.9, page 65; figures 8.19 and 8.20, pages 78 and 79; figure V.1, page 135; figure 18.1, page 136; figure 19.4, page 145; figure 21.2 [mittens], page 154; figure BM.2, page 173)

Patterns by Author

Heartwarming Mitts (figure 4.4, page 40; figure 5.1, page 47; figure 21.2, page 154), *Spin-Off*, Fall 2019
Selma Margau Sweater (figure 6.5, page 60), *Spin-Off*, Summer 2020
Cecilia's Bosom Friend Shawl (figure 19.4, page 145), *Spin-Off*, Spring 2022
Blanka sweater, unpublished (figure BM.2, page 173)

Designs by Others

Seguin, design by Quince (figure 7.1, page 61; figures 7.4–7.8, pages 63–64)
Walk Along, design by ANKESTRICK (figure 7.10, page 66; figure BM.1, page 156)
Gro Hat, design by Fiber Tales (figure 6.3, page 56; figure 21.2, page 154)
Veela, design by Libby Jonson (figure 9.6, page 88; figure 13.1, page 108)
Bag based on Field belt bag sewing pattern by Merchant & Mills (figure 6.3, page 56)
Two-End Knitted Mittens, pattern by Berit Westman in Tvåändsstickning (figure 17.1, page 130)
Nettle Shawl by Himalayan Allo Nettle Handicraft (figure 12.2, page 105; figure 14.1, page 114; figure 15.1, page 118)

APPENDIX 2: TOOLKIT

What tools and how many you need is a question of budget, space, and your spinning practice. Spinning can be anything from a hand-carved pushka with hand-teased wool in the pastures to an arsenal of spindles, wheels, and preparation tools.

I have lots of spindles of different kinds. Often, I get curious about a model or a maker and increase my spindle collection with perhaps a spindle a year. I use spindles both for quick exploration and larger projects. For spinning projects, I like to have a couple of spindles of the same type, but it is not necessary.

The first **spinning wheel** I bought and still use daily is my Kromski Symphony. The other three wheels featured in this book are also mine: a Majacraft Little Gem I use for traveling, a pre-production Kromski Mazurka I use mainly for spinning flax, and the great wheel for which I share custody with my friend Cecilia, who has the space for her. See appendix 1 for references to images.

I use two pairs of **hand cards**, both with leather pads. One is curved with a tight sett for fine wools, and one is straight with a medium sett for medium to coarse wools (for images, see appendix 1). Flat and curved are a matter of preference. The curved pair was made to order for me by a Finnish maker, by hand using traditional techniques and design. The flat ones were made in the 1980s and were also hand made with traditional techniques. I bought them second hand in 2024, and they had never been used. They are the best I have ever experienced. Most modern cards are made with a synthetic foam material in the cloth. If you are lucky, you can find used or antique cards with reasonably smooth and intact leather cloth, or a modern maker who uses leather. The sett of the teeth in cards is calculated as tpi (teeth per square inch). A tighter sett results in a higher tpi. Cards for medium to coarse wools usually have a tpi around 72. These work for most wools, but if you only card fine wools I would suggest a pair with a higher tpi.

In 2023, I met a spinner who had researched the making of hand cards. She had learned to make cards from the last card maker in Sweden, back in the 1980s. She told me that the best cards have their teeth placed diagonally across the pad. That way, she argued, the wool goes through more teeth on their way across the pad (without getting stuck through a too tight sett). When I look for used cards (which I do from time to time), I go only for the ones with diagonally placed teeth.

The weight of the cards is important—the heavier the cards, the more you need to work. I find that many antique cards are lighter than modern ones. They also seem to have different weights within the pair. If so, I make sure to use the heaviest one as my stationary card. I like cards that have a flat insertion of the handle into the paddle. This makes the card sleek and steady in my lap. The salmon tail construction in my 1980s cards has this design (see appendix 1 for image references), and I find it common in antique Swedish cards.

I like to use mini **combs** for the flow of the combing movement (or when I am out and about) and larger combs with a combing station for more of a production combing. The single-pitched combs will allow all fiber types to glide through the tines. I use these mainly for keeping coats together. The two-pitched (or multiple-pitched) pair will keep a firmer grip on the fibers, keeping the shorter undercoat in the comb when I pull the longer outercoat off. I use these when I want to separate fiber types.

The tines should ideally be (at least) as high as the row is wide. This way the comber will be able to use all the tines on the combs. If the tine setup is wider than the height of the tines, some will be unused and take up unnecessary space and weight. The tips of the tines need to be sharp enough to cut through the mass of the wool (and preferably blunt enough not to break skin). If they are too blunt, the comber needs to use more force. This may lead to broken fibers and/or strain for the comber. The spacing of the tines should be adapted to the wool you are combing—a wider spacing for coarser wools and narrower for finer wools.

When it comes to mini combs (that you hold one in each hand, as opposed to a combing station where you hold one comb with both hands), the weight of each comb is important. A single-pitched comb would therefore be lighter than a two-pitched comb. My single-pitched mini combs weigh between 3 and 3.8 ounces (between 90 and 110 grams) per comb. The heavier they are the more cumbersome the combing will be. An ergonomic handle and weight balance is also preferable.

If you plan to buy only one pair of combs, I recommend single-pitched (mini) combs, for reasons of ergonomics. They are lighter in weight than two-pitched combs and lighter to work with since there is less resistance.

For teasing, I use a **flicker**, mini combs, or the combing station. Sometimes my hands only. A flicker can seem costly for its sole purpose of opening up the staples. Wouldn't a dog brush work just as well? Well, it would. It would also break, several times in my experience. After a while you will have paid as much for x (plastic) dog brushes as you would have for one (wooden) flicker. Ask me how I know. I am also getting acquainted with a "Lock Pop" that I bought recently. It can be described as a table-mounted block with a carding pad; it is used in the same way you would a flicker but keeps both your hands free to hold the staple, with the table as a resistance.

Other Tools I Use

- A (Lazy) Kate to ply from. For spindle spinning, I also use a shoe box Kate (see description in chapter 10).
- A niddy-noddy to make skeins of a set circumference. I use this for making skeins from wheel-spun yarns. For spindle-spun yarns, I usually just use my leg—I place my ankle on the other knee in a figure four and wind the skein between my foot and my hand that is placed with the back against the knee. I learned this technique from a beautiful video of Navajo weaver Clara Sherman spinning on her Navajo spindle.
- A nostepinne for making the prettiest center-pull balls from my skeins. This works splendidly with the thumb as well, but perhaps not for singles. They tend to cut off the blood flow. I do own a skein winder and an umbrella swift, but I haven't used them for years. I generally place the skein around my knees when I wind a ball on my thumb.
- A small pin loom for 4 x 4 inch (10 x 10 centimeter) woven swatches. In chapter 6, you can see how I use a kitchen sponge for the same purpose.
- A magnifier (preferably with an integrated light) is a good idea for looking at details in the yarn. I also have a loupe with which I can get even closer.
- Scales for keeping track of skein weight. Sometimes I weigh piles of rolags or combed tops to have equal weight for the singles of one skein.
- Sturdy shelves for all my spinning books. Some of my favorites are *The Fleece & Fiber Sourcebook* by Deborah Robson and Carol Ekarius; *Secrets of Spinning, Weaving and Knitting in the Peruvian Highlands* by Nilda Callañaupa Alvarez; *The Spinner's Book of Yarn Designs* by Sarah Anderson; and *Respect the Spindle* by Abby Franquemont. Put this one on the shelf, too.

GLOSSARY

Allmoge breeds: The eleven Swedish heritage and conservation breeds, protected in gene banks.

Batt: Carded wool in a flat and airy shape, ready to spin as it is or to be rolled into a **rolag**.

Bird's nest: A combed **top** rolled into a spiral that looks like an egg in a bird's nest (figure 8.12, page 75).

Crimp: The waves of fibers (figure 1.3, page 7; figure 2.15, page 25).

Cut (butt) end: The end of the fibers that has been shorn off the sheep (figure 4.4, page 40).

Distaff: A stick to dress with wool for spinning on a spinning wheel or a spindle.

Dual (double) coat: A **fleece** that has **outercoat** and **undercoat** fibers that are significantly different from each other (figure 4.5, page 41).

Fiber hand (also back hand for wheel spinning): The hand that holds the fibers and controls the flow of fibers going to the spinning hand.

Fleece: The coat of wool on a sheep.

Follicle: The sack in the skin from which the fiber grows.

Full: To lightly felt a wool textile into a denser material through washing and agitation (figure 5.2, page 50).

Grist: The thickness of a yarn, often calculated in length per weight, for example, yards per pound.

Improved breed: See **modern breed**.

In the grease: Unwashed. If I spin in the grease, it means I spin wool that I haven't washed. Some people define "in the grease" as wool that still has **lanolin** left in it.

Jacketed: A sheep with a textile coat to protect the fleece.

Kemp: Coarse, brittle fiber with a large and air-filled core (figure 1.2, page 6; figure 2.17, page 26).

Landrace: A breed specific to a local area.

Lanolin: Wool grease produced in glands in the sheep's skin.

Long draw: Letting the twist into the undrafted fibers during the draft (figure 9.6, page 88; figure 14.2, page 115; figure 20.2, page 149).

Micron count: A measurement in micrometers to determine fiber fineness.

Modern breed: A sheep breed that has been bred in modern times for the benefit of specific characteristics (and/or to the detriment of others).

Moulting: The capacity in some **primitive breeds** to shed their fleece.

Niddy-noddy: A hand tool for winding skeins.

On the hoof: The fleece on the body of the sheep.

Outercoat: Straight, smooth, long, and strong hair fibers in a **fleece** (figure 4.5, page 41).

Park and draft: Spindle spinning where the spinner stops and parks (or stops the rotation of) the spindle to separate the actions in the spinning.

Pick: To draw out single **staples** of a **fleece** to open it up (figure 4.1, page 39).

Polled: Hornless.

Primitive breed: A breed with primitive characteristics like horns, **dual coats**, and **moulting**, characteristics that often have been bred off in **modern breeds**.

Raw fleece: Unwashed fleece.

Rise: A spot on some **primitive breeds** where one year's growth thins out and may or may not break when the next year's growth emerges.

Rolag: A carded **batt** rolled into a cylinder (figure 8.2, page 69).

Scour: Wash thoroughly to remove all dirt and grease.

Second cuts: Short pieces of fiber where the shearer has two cutting actions in a single section of the fleece.

Semi-woolen: A yarn in a **woolen** preparation spun with typically **worsted** spinning techniques.

Semi-worsted: A yarn in a **worsted** preparation spun with typically **woolen** spinning techniques.

Shedding: *See* **moulting**.

Short draw: Drafting the fibers before letting the twist in (figure 10.1, page 91).

Slub: A bump or thick spot in a yarn.

Spinning from the fold: Spinning sections of fibers folded over a finger (figures 10.3 and 10.4, page 94).

Spinning hand (also front hand for wheel spinning or spindle hand for spindle spinning): The hand that holds the yarn and controls the twist moving into the fibers.

Spinning triangle: The triangle shape created where fibers and twist meet (figure 9.1, page 85).

Staple: A cluster of adjacent fibers joined by a combination of characteristics, for example, length, crimp, and speed of growth (figure 2.2, page 13).

S-twist/S-ply: A yarn spun counterclockwise is said to have an S-twist; the angle of the twist follows the direction of the middle part of the letter *S*. Similarly, a yarn plied counterclockwise is said to be S-plied.

Suint: A sheep's sweat.

Tease: To lightly open up locks (figure 7.2, page 62).

Tip end: The outer (free) end of a **fleece** (figure 4.4, page 40).

Top: Combed fibers that have been aligned and drawn out into a long stretch of fibers (figure 8.18, page 78).

Twist angle: The degree of twist in a yarn.

Undercoat: Fine, elastic, and often crimpy wool fibers in a **fleece** (figure 8.20, page 79).

Vegetable matter (sometimes vegetation matter, aka "vm"): Pieces of nature (seeds, grass, twigs), food, or bedding (hay, straw) in a **fleece**.

Waulk: To (heavily) **full**.

Whorl: A weight on a spindle.

Woolen spinning: Spinning, usually from carded wool, with the aim of a soft and airy yarn (figure 10.2, page 92).

Worsted spinning: Spinning, usually from combed wool, with the aim of a strong and shiny yarn (figure 10.1, page 91).

Z-twist/Z-ply: A yarn spun clockwise is said to have a Z-twist; the angle of the twist follows the direction of the middle part of the letter *Z*. Similarly, a yarn plied clockwise is said to be Z-plied.

BIBLIOGRAPHY

Prologue

Kimmerer, Robin Wall. *Braiding Sweetgrass: Indigenous Wisdom, Scientific Knowledge, and the Teaching of Plants.* Minneapolis, MN: Milkweed Editions, 2013.

Chapter 1: Off the Hoof: Listen to the Sheep

American Wool Council. "Wool Grades and the Sheep That Grow the Wool." American Sheep Industry Association, June 2022. Accessed January 6, 2024. https://www.sheepusa.org/wp-content/uploads/2022/06/Wool_Grades_and_the_Sheep_that_Grow_the_Wool_Scan-1.pdf.

Animalia. "Norwegian Wool Standard." Landsbygdsdirektoratet, 2022. Accessed January 5, 2024. https://www.animalia.no/contentassets/d91150be325e4d72b5b814f83b92b2f8/202961-animalia-ullstandard-engelsk-04.pdf.

——. "Ulltyper." January 23, 2023. Accessed January 6, 2024. https://www.animalia.no/no/Dyr/ull-og-ullklassifisering/ulltyper/.

AWEX. "AWEX–Sheep Breeds." Australian Wool Exchange, July 2020. Accessed January 7, 2024. https://www.sasheepexpo.com.au/wp-content/uploads/2020/07/Sheep-Breeds-AWEX-SA_t.pdf.

Axfoundation and Arena svensk ull. "Svensk ullstandard." Axfoundation, 2025. Accessed April 9, 2025. https://www.axfoundation.se/svensk-ullstandard.

Beck, Lena. "'Waste Wool' Is a Burden for Farmers: What If It Could Be a Solution Instead?" Modern Farmer, November 6, 2023. Accessed January 6, 2024. https://modernfarmer.com/2023/11/waste-wool-solution/.

British Wool. "British Sheep and Wool." British Wool Marketing Board, 2010. Accessed January 6, 2024. https://www.britishwool.org.uk/assets/brochures/Breed-Book.pdf?fbclid=IwAR2wNXNpoglbphInAlu-mIYepSNOj79Ke9Dqq7DZKI1HpQQ_6sSgGiSKONw (article is no longer available).

Coulthard, Sally. *A Short History of the World According to Sheep.* London: Head of Zeus, 2021.

Davidson, Gordon. "What a Waste of Wool!" Scottish Farmer, 2021. Accessed January 6, 2024. https://www.thescottishfarmer.co.uk/news/19509147.waste-wool/.

Hult, Annkristin, et al. *Om ull.* Stockholm: Svenska hemslöjdsföreningarnas riksförbund, 2022.

Jordbruksverket. "Kategorisering av animaliska biprodukter." 2015. Accessed January 7, 2024. https://jordbruksverket.se/download/18.71b6fea1174ae10f764d40c9/1600853094662/Kategorisering-av-animaliska-biprodukter.pdf.

Kristiansen, Kristian, and Marie Louise Stig Sørensen. "Wool in the Bronze Age: Concluding Reflections." In *The Textile Revolution in Bronze Age Europe: Production, Specialization, Consumption,* edited by Serena Sabatini and Sophie Bergerbrandt, 317–32. Cambridge: Cambridge University Press, 2020.

Martiin, Carin, ed. *Den svenska fårskötseln äldre historia: Några kapitel ur Ull och Ylle–av Sven T. Kjellberg.* Stockholm: Kungl. Skogs- och lantbruksakademien, 2009.

Robson, Deborah, and Carol Ekarius. *The Fleece & Fiber Sourcebook: More Than 200 Fibers from Animal to Spun Yarn.* North Adams, MA: Storey, 2011.

Ryder, Michael L. *Sheep and Man.* Norwich: Duckworth, 1983.

Sabatini, Serena, and Sophie Bergerbrandt. "Textile Production and Specialization in Bronze Age Europe." In *The Textile Revolution in Bronze Age Europe: Production, Specialization, Consumption,* edited by Serena Sabatini and Sophie Bergerbrandt, 1–14. Cambridge: Cambridge University Press, 2020.

Skals, Irene. "To Let Textiles Talk: Fibre Identification and Technological Analyses of Prehistoric Textiles from Denmark." In *The Textile Revolution in Bronze Age Europe: Production, Specialization, Consumption,* edited by Serena Sabatini and Sophie Bergerbrandt, 134–53. Cambridge: Cambridge University Press, 2020.

Söderberg, Fia. "Om oss." Ullförmedlingen. Accessed January 7, 2024. https://ullformedlingen.se/om-ullformedlingen.

Svenska fåravelsförbundet. "Färsk statistik om den svenska ullen." May 10, 2021. Accessed January 4, 2024. https://faravelsforbundet.se/farsk-statistik-om-den-svenska-ullen/.

Ullkontoret. "Sveriges enda storskaliga ulltvätteri." Accessed January 7, 2024. https://ullkontoret.se/ULLKONTORET/Om_tvatteriet.html.

Vitlycke Museum. "Bronsålderns ullekonomi." Facebook, 2022. Accessed December 14, 2023. https://www.facebook.com/vitlyckemuseum/videos/845880379954335.

Woola. "Why Wool? And Why Do Farmers Burn It?" 2023. Accessed January 6, 2024. https://www.woola.io/blog/why-wool.

Chapter 2: A Smorgasbord of Wool

Avelsföreningen jämtlandsfåret. "Jämtlandsfåret." Facebook, August 29, 2016. Accessed December 2, 2023. https://www.facebook.com/permalink.php?story_fbid=pfbid0bvz4PM7iVPhsup6qnCr3BAe9cJ23fYj7pj7MpzAEo8jdnsxtuEWnEwdjhLB5Ezd4l&id=677743648942195.

Föreningen ryafåret. "Föreningen Ryafåret—för ryafåret i framtiden." 2023. Accessed November 24, 2023. https://www.ryafaret.se/.

Föreningen svenska allmogefår. "Föreningen svenska allmogefår." 2023. Accessed December 3, 2023.https://www.allmogefar.se/.

Föreningen svenska finullsfår. "Föreningen svenska finullsfår." 2023. Accessed November 28, 2023. http://finull.se/.

Gotlandsfårföreningen. "Gotlansfårföreningen." 2023. Accessed December 2, 2023. http://www.silverlock.se/.

Gustafsson, Kerstin, and Alan Waller. *Ull: Hemligheter, möjligheter, färdigheter.* Helsingborg, Sweden: Läs förlag, 1987.

Hult, Annkristin, et al. *Om ull.* Stockholm: Svenska hemslöjdsföreningarnas riksförbund, 2022.

Parholt, Anntott, et al. *Nock, ragg, rya—det glänser om ullen.* Örebro, Sweden: Föreningen Sveriges hemslöjdskonsulenter UllMa, 2001.

Sjödin, Erik, et al. *Får.* Stockholm: Natur och Kultur, 2011.

Söderberg, Fia. "Ryafåret—hotat får med potential." Ryafåret, 2018. Accessed November 23, 2023. https://www.ryafaret.se/wp-content/uploads/2018/06/ryaartikel-1805.pdf.

Svenska fåravelsförbundet. "Elitlamms årsstatistik för 2023." 2024. Accessed August 18, 2024. https://faravelsforbundet.se/elitlamms-arsstatistik-for-2023/.

Svenska leicesterfårföreningen (website). 2023. Accessed December 3, 2023. http://www.leicesterfarforeningen.se/.

Swedish Poll Merino Association. "Swedish Poll Merino Association." 2023. Accessed December 10, 2023. https://svenskmerino.webnode.se/.

Chapter 3: On Washing Fleece

Alvarez, Nilda Callañaupa, and the Weavers of the Center for Traditional Textiles of Cusco. *Secrets of Spinning, Weaving and Knitting in the Peruvian Highlands.* Loveland, CO: Thrums Books, 2017.

Martiin, Carin, ed. *Den svenska fårskötseln äldre historia: Några kapitel ur Ull och Ylle—av Sven T. Kjellberg.* Stockholm: Kungl. Skogs- och lantbruksakademien, 2009.

Merrow, Anne. "Three Spinners on Washing Wool." *Spin-Off,* February 23, 2022. Accessed January 21, 2024. https://spinoffmagazine.com/spinners-washing-wool/.

——. "Wool Washing." *Spin-Off,* July 11, 2017. Accessed January 21, 2024. https://spinoffmagazine.com/washing-wool-free-guide/.

Ryder, Michael L. *Sheep and Man.* Norwich: Duckworth, 1983.

Schröder-Gravendyck, A. Sabine. "How to Wash Fleece and Leave the Grease." *Spin-Off,* August 17, 2020. Accessed January 27, 2024. https://spinoffmagazine.com/how-to-wash-fleece-and-leave-the-grease/.

——. "Wool Basics: What Is Grease?" *Spin-Off,* August 29, 2022. Accessed January 21, 2024. https://spinoffmagazine.com/wool-basics-what-is-grease/.

Chapter 4: Staple by Staple: Pick Your Fleece

Anderson, Enid. *The Spinner's Encyclopedia.* Devon: David & Charles, 1987.

Ross, Mabel. *Encyclopedia of Handspinning*. London: Batsford, 1988.

Chapter 6: Sampling, Swatching, and Keeping Records

Waltin, Josefin. "Slow Fashion 2—from Sheep to Shawl." YouTube, 2018. Accessed January 27, 2025. https://www.youtube.com/watch?v=SKYXUayGxWE&t=283s.

Chapter 7: Add Air: Tease Your Wool

Alvarez, Nilda Callañaupa, and the Weavers of the Center for Traditional Textiles of Cusco. *Secrets of Spinning, Weaving and Knitting in the Peruvian Highlands*. Loveland, CO: Thrums Books, 2017.

Gustafsson, Kerstin, and Alan Waller. *Ull: Hemligheter, möjligheter, färdigheter*. Helsingborg: Läs förlag, 1987.

Siska, Maja. "Spin It! Lopi-Style Yarn." *PLY* 32 (Spring 2021).

Chapter 8: Prepare: Shape the Mass and Structure the Fibers with Cards and Combs

Demers, Carson. *Knitting Comfortably. The Ergonomics of Handknitting*. San Francisco, CA: Ergo Publishing, 2016.

Long Thread Media. "How to Card Wool: Four Spinners, Four Techniques." Accessed August 29, 2024. https://shop.longthreadmedia.com/products/how-to-card-wool-download-in-hd.

Robson, Deborah. "Primitive and Double-Coated." *PLY* 32 (Spring 2021).

Chapter 9: Twist Model

Waltin, Josefin. "The Flick." *PLY* 29 (Summer 2020).

Chapter 10: Spin, Ply, and Finish

Condon, Meagan. "Woolen to Worsted." *PLY* 30 (Autumn 2020).

Chapter 11: Be Kind

Demers, Carson. *Knitting Comfortably: The Ergonomics of Handknitting*. San Francisco, CA: Ergo, 2016.

Friskis & Svettis. "Friskismotcorona." Instagram, 2020. Accessed May 8, 2024. https://www.instagram.com/p/B91CknkhDh-/?igsh=MXFucmY4c25mOTF5aw==.

Lundberg, Göran. *Trepunktnoll—Handen. Hjärnan. Tiden*. Stockholm: Carlssons bokförlag, 2022.

Merriam, Sharan B., and Laura L. Bierema. *Adult Learning: Linking Theory and Practice*. San Francisco, CA: Jossey-Bass, 2014.

Chapter 14: Closeness

Barber, Elizabeth Wayland. *Women's Work: The First 20,000 Years: Women, Cloth and Society in Early Times*. New York: Norton, 1994.

Lundberg, Göran. *Trepunktnoll—Handen. Hjärnan. Tiden*. Stockholm: Carlssons bokförlag, 2022.

Chapter 15: Slow

Alvarez, Nilda Callañaupa, and the Weavers of the Center for Traditional Textiles of Cusco. *Secrets of Spinning, Weaving and Knitting in the Peruvian Highlands*. Loveland, CO: Thrums Books, 2017.

Chapter 16: We Need to Talk

Merriam, Sharan B., and Laura L. Bierema. *Adult Learning: Linking Theory and Practice*. San Francisco, CA: Jossey-Bass, 2014.

Chapter 17: Embrace Your Mistakes

Kempton, Beth. *Wabi Sabi: Japanese Wisdom for a Perfectly Imperfect Life*. London: Piatkus, 2018.

Kimmerer, Robin Wall. *Gathering Moss: A Natural and Cultural History of Mosses*. Dublin: Penguin Random House, 2021.

Waltin, Josefin. "Heartwarming Mitts." *Spin-Off*, Autumn 2019.

———. "Twist Analysis in Twined Knitting." *Spin-Off*, Autumn 2019.

Chapter 18: The Knowledge of the Hands

Lundberg, Göran. *Trepunktnoll—Handen. Hjärnan. Tiden*. Stockholm: Carlssons bokförlag, 2022.

Chapter 19: With My Two Hands

Bojsen-Møller, Finn. *Rörelseapparatens anatomi*. Stockholm: Liber, 2023.

Lundberg, Göran. *Trepunktnoll—Handen. Hjärnan. Tiden*. Stockholm: Carlssons bokförlag, 2022.

Waltin, Josefin. "Cecilia's Bosom Friend." *Spin-Off*, Spring 2022.

Chapter 20: The Inner Spinner

Lundberg, Göran. *Trepunktnoll–Handen. Hjärnan. Tiden.* Stockholm: Carlssons bokförlag, 2022.

Chapter 21: Let the Wool Be Your Teacher

Ende, Michael. *Momo–eller kampen om tiden.* Malmö, Sweden: Berghs förlag AB, 1980.

Appendix 2: Toolkit

Wolf Creek. "Navajo Weaver Clara Sherman Carding and Spinning." YouTube, 2010. Accessed January 31, 2025. https://www.youtube.com/watch?v=D_p7OIghMVw.

INDEX

J

K

L

M

N

O

P

R

S

T

U

V

W

Y

ABOUT THE AUTHOR

Josefin Waltin is a spinner, spinning teacher, writer, and course creator. She has written several articles for magazines like *Spin Off* and *PLY* and is the coauthor of Sara J Wolf's book *Knit (Spin) Sweden!* Her knowledge of wool comes straight from the source—from the raw fleece itself. With her hands in the wool and by writing about spinning, she understands the process and deepens her connection to the wool. She has taught more than three hundred hours in face-to-face courses and more than two thousand students online. As a spinning teacher and course creator, she has collected a bank of experiences and common challenges for spinners and brought it back into teaching and writing. Her main focus is spindle spinning and the process from raw fleece to a finished yarn or textile. Josefin lives in Stockholm, Sweden, with her family. She gardens, is a trained gym instructor, and enjoys daily swims all year round in her nearby lake.

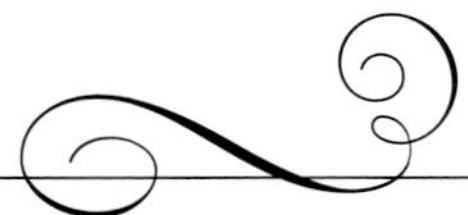

You can find Josefin here:
Substack: https://substack.com/@josefinwaltin
Instagram: https://www.instagram.com/josefinwaltin
Blog: https://waltin.se/josefinwaltinspinner
Online spinning school: https://josefinwaltinspinner.teachable.com
YouTube: https://www.youtube.com/@josefinwaltin

BM.2 In the pasture with curious Värmland and Gestrike sheep.